BIG COAL

GUY PEARSE is a former lobbyist and Canberra insider who now writes mainly on energy and environment issues, with an emphasis on corporate and political collusion, deception and spin. He is the author of *Greenwash: Big brands and carbon scams*, *High & Dry: John Howard, climate change and the selling of Australia's future* and the Quarterly Essay *QuarryVision: Coal, climate change and the end of the resources boom*.

Politics and media commentator DAVID McKNIGHT is the author of *Rupert Murdoch: An investigation of political power*, as well as *Beyond Right and Left: New politics and the culture wars* and *Australia's Spies and their Secrets*. He has worked as a journalist on the *Sydney Morning Herald*, ABC TV's *Four Corners* and the weekly *Tribune* and is a senior research fellow at the University of New South Wales.

BOB BURTON is an environment writer and author of *Inside Spin: The dark underbelly of the PR industry* and co-author of *Secrets and Lies: The anatomy of an anti-environment PR campaign*. He has worked for a range of environmental groups and is contributing editor of the CoalSwarm wiki and a director of the Sunrise Project, a group which advocates a transition away from fossil fuels.

BIG COAL
AUSTRALIA'S DIRTIEST HABIT

GUY PEARSE   DAVID McKNIGHT   BOB BURTON

NEWSOUTH

A NewSouth book

Published by
NewSouth Publishing
University of New South Wales Press Ltd
University of New South Wales
Sydney NSW 2052
AUSTRALIA
newsouthpublishing.com

© Guy Pearse, David McKnight, Bob Burton 2013
First published 2013

10 9 8 7 6 5 4 3 2 1

This book is copyright. Apart from any fair dealing for the purpose of private study, research, criticism or review, as permitted under the Copyright Act, no part of this book may be reproduced by any process without written permission. Inquiries should be addressed to the publisher.

National Library of Australia
Cataloguing-in-Publication entry
Author: Pearse, Guy.
Title: Big coal: Australia's dirtiest habit/Guy Pearse,
    David McKnight, Bob Burton.
ISBN: 9781742233031 (paperback)
    9781742241463 (epub/mobi)
    9781742246406 (epdf)
Subjects: Coal trade – Australia.
        Coal – Environmental aspects – Australia.
        Coal trade – Economic aspects – Australia.
        Exports – Government policy – Australia.
Other Authors/Contributors: McKnight, David, author.
    Burton, Bob, 1959– author.
Dewey Number: 338.2724

Design Di Quick
Cover design Nada Backovic
Cover images Corbis

All reasonable efforts were taken to obtain permission to use copyright material reproduced in this book, but in some cases copyright could not be traced. The author welcomes information in this regard.

CONTENTS

Abbreviations vii

Foreword by *Ian Dunlop* ix

Introduction 1

1 The real price of coal 12

2 The strip-mining of Australia 34

3 The barons' boom 60

4 Charm offensive 100

5 King Coal's muscles 128

6 'Clean coal' ruse 156

7 Overthrowing Big Coal 194

Notes 220

Index 247

To our children

ABBREVIATIONS

ABARES	Australian Bureau of Agricultural and Resource Economics and Sciences
ACA	Australian Coal Association
ACARP	Australian Coal Association Research Program
ACSMP	Australian Centre for Sustainable Mining Practices
AGL	Australian Gas Light
AIGN	Australian Industry Greenhouse Network
AMEC	Association of Mining and Exploration Companies
ANDEV	Australians for Northern Development and Economic Vision
ASX	Australian Stock Exchange
BICC	Blackwater International Coal Centre
BMA	BHP Mitsubishi Alliance
BNP Paribas	Bank Nationale de Paris/Paribas (merged)
BREE	Bureau of Resources and Energy Economics
CCR	carbon capture and recycling
CCS	carbon capture and storage
CCSG	Centre for Coal Seam Gas
CFMEU	Construction, Forestry, Mining and Energy Union
CIAB	Coal Industry Advisory Board
CIC	China Investment Corporation
CITIC	China International Trust and Investment Corporation
CITIC	Pacific Mining
CLET	Centre for Low Emissions Technology
CO_2CRC	CO_2 Cooperative Research Centre
CSIRO	Commonwealth Scientific and Industrial Research Organisation
CSLF	Carbon Sequestration Leadership Forum
DEEDI	Queensland Department of Employment, Economic Development and Innovation

DIDO	drive in, drive out
EDO	NSW Environmental Defenders Office
ENGO Network on CCS	Environmental NGO Network on Carbon Capture and Sequestration
ESAA	Energy Supply Association of Australia
FIFO	fly in, fly out
FMG Resources	Fortescue Metals Group
GCCSI	Global Carbon Capture and Storage Institute
GIC	Government of Singapore Investment Corporation Private Limited
GVK	Indian conglomerate, G.V.K Reddy, owner
HRL	formerly Hydrometallurgy Research Laboratories
HSBC	The Hong Kong and Shanghai Banking Corporation
ICAC	Independent Commission Against Corruption (NSW)
IEA	International Energy Agency
IGCC	Integrated Gasification Combined Cycle
IMF	International Monetary Fund
IPA	Institute of Public Affairs
IPCC	Intergovernmental Panel on Climate Change
JFE Steel Corporation	Jeiefui Hōrudingusu Kabushiki gaisha
KIA	Kuwait Investment Authority
LETAG	Lower Emissions Technology Advisory Group
MCA	Minerals Council of Australia
MRET	mandatory renewable energy target
RBS	Royal Bank of Scotland
SGH	Seven Group Holdings
SMI	Sustainable Minerals Institute
UBS	originally Union Bank of Switzerland, now merged with Swiss Bank Corporation
UCG	underground coal gasification technology
UNESCO	United Nations Educational, Scientific and Cultural Organisation
WHO	World Health Organisation

FOREWORD
BY IAN DUNLOP

Australia is teetering on the brink of the greatest strategic blunder in its history. If planned expansion of the coal industry proceeds, Australia will find itself 'beautifully equipped for a world which no longer exists', with extensive stranded assets in mines, ports and railways, as key trading partners like China and India rapidly abandon a high-carbon future in favour of low-carbon alternatives.

Coal has a long and turbulent history. It was the mainspring of the Industrial Revolution. Its cheap energy subsequently leveraged the creation of phenomenal wealth in the developed world, as well as much of the wealth being created today in the developing world. Along the way, it has been the crucible for some of the most profound social conflict and reform in history, in the perennial confrontation between capital and labour. The harshness of working conditions, and mutual dependency in underground mining between both miners themselves, and between miners and management, bred an incredibly socially cohesive industry, with a strong sense of justice, which could be hugely constructive when pursuing common goals, or hugely destructive otherwise.

In the four decades after the Second World War, coal, along with other mining development, was at the forefront of Australia's growth, guided by industry leaders who were statesmen in the true sense of the word, intent upon creating not just profitable enterprises, but a long-term bedrock for national prosperity. As Japan restructured and other Asian economies began to move up the growth escalator, Australia benefited greatly from establishing long-term relationships and industries to meet their demands.

The coal industry itself became ever more professional and technically proficient, to the point that today, at an operational level, it is without doubt the world leader in coal mining and development.

Sadly, the same cannot be said for the industry's leaders who, in recent times have lost their way. Over the last two decades, the accumulation of excessive economic and political power, combined with performance-related remuneration for senior executives, has led to the dominance of short-term thinking and decision making. Statesmanship has disappeared, along with any thoughts of sustainable nation-building, and it has been replaced with an overriding emphasis on immediate financial gain.

Simultaneously, coal's great nemesis, climate change, has become a reality. The industry has been well aware that carbon emissions from coal consumption would, at some stage, become a major constraint on its future, having researched the issue in depth since the 1980s. Every major coal company, in its corporate responsibility and sustainability policies, acknowledges that climate change is a serious issue that needs to be urgently addressed. In reality, those same companies, via proxies such as the national and state Minerals Councils and the Australian Coal Association, have fought tooth and nail to prevent the introduction of sensible climate change policy in Australia.

It is now evident that climate change is accelerating far faster than previously anticipated, and that increasing coal consumption is a major cause. 'Official' solutions, such as carbon capture and storage and other clean coal technologies, are clearly not working and will not deliver carbon emissions reductions, either to the extent or in the limited time now required. Despite this, the industry proposes to double coal exports by 2025 with no means of sequestering the carbon.

Current climate change policies, such as the federal government's 'Clean Energy Futures' package, will deliver a global temperature increase of at least 4°C, probably more, rather than the 'official' target of 2°C. This would lead to the death of 6 billion people, leaving a world of 1 billion rather than the current 7 billion people, with horrific implications for our children and grandchildren. What could have been sensible policy was corrupted by coal industry lobbying to produce this disastrous outcome.

Ross Garnaut made the point in his 2008 *Climate Change Review*, that: 'The most costly and damaging policy for Australia would be to implement a policy that was designed to appear meaningful, but was largely meaningless in application'. That is exactly what the coal industry has delivered.

John Maynard Keynes asked: 'When the facts change, I change my mind. What do you do, Sir?' The facts for the coal industry have now changed, and yet there are no coal industry leaders prepared to change their minds, publicly acknowledge that the industry is no longer sustainable and that it has to be phased out.

Unless this happens rapidly, irreversible, catastrophic outcomes will become locked in. The proponents of expansion are largely individuals with no experience of coal, its social history or the development of major mining projects. They are in total denial of climate change and such implications, oblivious to the fact that every new fossil-fuel project represents death and destruction for communities somewhere in the world, Australia included. Given that our political and business leaders are well aware of the extreme climatic risks we now run, in promoting coal they are wilfully perpetuating nothing less than a crime against humanity.

We have solutions to climate change. Australia has enormous ingenuity, low-carbon resources and opportunities, but only if we are honest about the real challenge and take emergency action to change our current unsustainable direction.

What is good for coal is no longer automatically good for Australia. The coal industry's social and financial licences to operate are in the process of being withdrawn. It is time for coal industry leaders to put the attack dogs of the Australian Coal Association and Minerals Councils back in their kennels and intelligently redirect the industry's management, professional and financial skills towards developing low-carbon alternatives. New technologies may one day emerge that will open up other opportunities for coal, but that is another story.

In this book, Guy Pearse, David McKnight and Bob Burton have done great service in clearly and objectively explaining how coal has come to dominate policy formulation in Australia, why it has become such a danger to our democracy, and how that domination is being broken. It is essential reading to anyone interested in practical solutions to break the climate change impasse and create a genuinely sustainable future for Australia. I commend it to you.

||||

Ian Dunlop is a former international oil, gas and coal industry executive. He chaired the Australian Coal Association in 1987–88, chaired the Australian Greenhouse Office Experts Group on Emissions Trading from 1998 to 2000 and was CEO of the Australian Institute of Company Directors from 1997 to 2001. He is a Director of Australia21, Chairman of Safe Climate Australia, a Fellow of the Centre for Policy Development and a Member of the Club of Rome.

INTRODUCTION

Grazier and landowner Paola Cassoni lives in central-western Queensland. Back in 2000 she was worried about the extensive and rapid clearing of the land taking place around her. Along with other local families, she helped purchase 8000 hectares of land that was once part of an old cattle property near the small town of Alpha. Together they aimed to preserve the remnant woodland, along with its birds, reptiles and other animals. With the financial support of the federal government the land was turned into a nature reserve called Bimblebox Nature Reserve and became part of the National Reserve System with the agreement of the state government. Paola Cassoni and the other owners were happy to sign a legally binding agreement stating that it would be conserved forever.[1]

Paola Cassoni thought this guaranteed that the land would be protected. But she did not reckon on the global coal boom about to take place, nor on the state government's willingness to green light almost any mine, anywhere, apparently without hestitation. By 2008 it had become clear that much of the Bimblebox reserve might be swallowed up by the China First coal project, owned by mining magnate Clive Palmer. According to the Environment Impact Statement drawn up by Palmer's Waratah Coal, half the Bimblebox Nature

Reserve would be affected by the open-cut mine while the other half would suffer from subsidence caused by underground mining.[2] The land owners were shocked. For his part, Clive Palmer rejected their concerns. 'There's nothing there of any environmental significance. If people want to dress up as kangaroos and koalas, they can do it,' he said.[3] Yet the Bimblebox Nature Reserve was only designated as such after it met globally agreed standards of biological diversity.

Paola Cassoni feels betrayed by the Queensland state government, which she accuses of 'shamefully abandoning the conservation covenants they signed with landholders'.[4] She has vowed to fight the mining plan and has gone on speaking tours around the state, helped to make a film about the issue and linked up with others who are protesting against the radical expansion of gas and coal mining across Australia.

At the other end of Australia a very different protest took place. It happened on 9 June 2010 in Perth, Western Australia, when a rally of 2000 people assembled at Langley Park. These protesters were not worried about wildlife, woodland or nature reserves. Prime Minister Kevin Rudd was in town to defend his tax on the super profits of the mining industry. The protesters intended to give him a piece of their minds. Yet the rally was a little odd. It was organised by mining companies. Many protesters had been bussed in by their employers. On arrival they were handed neatly printed placards to wave. Many read 'Rudd's mining tax hurts us all', echoing the theme of a $22 million advertising campaign that had been blitzing the airwaves for four weeks.

At the front of the rally, on the back of a flat-bed truck, were Australia's second and fourth richest people – Gina Rinehart ($4.7 billion) and Andrew Forrest ($4.2 billion). It was 'billionaires at the barricades', said a journalist later.[5] During the rally one mining company executive rang a bell and shouted: 'I'm Australian and I love Australia. You are robbing the elderly, the sick and the frail'.[6] Wearing a string of pearls, the reclusive Gina Rinehart bellowed: 'What are we gonna tell those jittery Labor MPs in marginal seats?'[7] The crowd responded with a chant, 'Axe the tax! Axe the tax!' As she continued to lead the chanting, Gina Rinehart's voice became hoarse. Later she explained, 'The reason I am, you know, screaming up and down about this, even in city streets, is because I feel so strongly about it. I feel strongly about it for our future of this country.'[8]

Gina Rinehart's impact on the future of Australia will be profound. Along with other coal miners she hopes to open up the vast Galilee coalfields in Central Queensland, which cover the nature reserve that Paola Cassoni cares so deeply about.

The radical expansion of coal mining, of which this is part, will have increasingly damaging effects on Australia. It will affect not only nature reserves and taxes, but also icons like the Great Barrier Reef. Nine new or expanded ports are proposed and many require extensive dredging of areas on the Queensland coast. In some areas this dredging is already killing fish and damaging the habitat within which they survive. The export boom in coal will turn the ocean around the reef into a coal super highway. While the quadrupling of ships will make collisions and wrecks more likely, this is not the main threat. When the exported coal is eventually burned, emitting greenhouse gases, this will unavoidably damage the reef. This was the fear expressed in a statement by 2000 marine scientists who met in Cairns in July 2012, warning about the warming and acidification of the world's oceans. 'This combined change in temperature and ocean chemistry has not occurred since the last reef crisis 55 million years ago,' they said.[9]

And it's not just the Barrier Reef. The growing accumulation of greenhouse gases in the atmosphere and the resultant warming is already worsening Australia's climate. Over several weeks in January 2013 Australia experienced its worst heatwave. A dome of heat settled over Australia and refused to move. The monsoon, which usually brings cloud and rain to Australia's north, remained stubbornly offshore, while parts of the continent turned into a furnace. Overall, the national daily average maximum of 40.33°C broke the previous record set in 1972. According to the Weather Bureau's manager of climate monitoring, David Jones, '...the climate system is responding to the background warming trend. Everything that happens in the climate system now is taking place on a planet which is a degree hotter than it used to be.'[10] Australia's Climate Commission agreed. In a report titled *Off the Charts*, one of its authors, Professor David Karoly, said that, 'Although Australia has always had heatwaves, hot days and bushfires, climate change is increasing the risk of more frequent and longer heatwaves and more extreme hot days, as well as exacerbating bushfire conditions'.[11]

Yet for most Australians the shiny black lumps of coal that we dig up or burn are remote from our daily lives. If we think about coal at all, we dimly realise that our laptops, electric stoves and lights depend on the electricity produced by burning this ancient rock. Occasionally, we see a news broadcast or an advertisement trumpeting Australia's growing coal exports to an energy-hungry world. While coal was once burned in our fireplaces and in city-based power stations, today most of us don't smell or touch it. It is scooped up from distant mines, loaded onto trains and disappears into comfortably remote power stations via ships that sail off over the horizon. Out of sight, out of mind. A comforting illusion is fostered that coal is benign.

However, it's time Australians thought much more about coal. Our future is increasingly tied up in it. To generate electricity Australia burns proportionally more coal than other advanced industrial countries – about 75 per cent of our electricity is generated from coal-fired plants. Overall, more than 40 per cent of our primary energy supply comes from coal, twice as much as comparable industrialised countries.[12] Australian export coal constitutes one-third of the world's seaborne coal trade. This global coal trade is increasingly feeding the power plants and steel mills of India and China, both responsible for much of the increasing global greenhouse emissions.

At home, the coal industry hopes to double Australia's coal exports in the next decade or two. Some claim that the 'boom' is over, but that simply means the price of coal has dropped and some new projects have been postponed. Coal remains profitable and attractive, especially if it can be exported in higher volumes. A radical expansion of coal mining is under way and Australians are encouraged to celebrate this growth and not to ask questions. But questions need to be asked. Fundamentally, we need to ask whether our increasing dependence on mineral exports, especially coal, is a road to prosperity for Australia or a dead end?

On the surface, coal is a $48 billion export industry, directly employing around 46 000 people and attracting huge investment. Probe a little deeper and it's not so impressive. For starters, the mining industry is around 80 per cent foreign-owned and therefore most profits eventually go offshore.[13] Mining companies, such as Xstrata, Peabody and Anglo American, are 100 per cent foreign owned, with others overwhelmingly in overseas hands (Rio Tinto 83 per cent, BHP 76 per cent). The industry claims that much is spent on employing people. In reality, a tiny share of this money is spent on payroll as coal has become increasingly mechanised. Coal directly employs 46 000 people, but this is tiny in an Australian workforce of over 11 million.[14] At the height of the recent coal boom in 2012, the jobs justification was fatally weakened when the big Alpha mine (jointly owned by the Indian company GVK and Gina Rinehart) admitted they were prepared to import foreign workers if needed to build the mine. Similar statements have been made by mining magnate Clive Palmer and the Indian company Adani, both of whom plan to export coal from massive mines in the Galilee Basin in Queensland.[15]

As for the investment by coal companies, nearly all of this money is spent on equipment, mining camps, railways and ports for the near-exclusive use of the coal industry. The wider community enjoys few of the benefits from these investments. The expansion of coal mining is damaging other export industries such as agriculture and education. Mining has driven up the value of the Australian dollar, so Australian wheat, wool and manufactured exports are more expensive to overseas customers. A higher dollar jacks up the cost of fees paid by overseas students to study in Australia and makes it more expensive for tourists to visit here. Local tourism is also hit because the high dollar makes it cheaper for Australians to take holidays overseas. The high dollar makes manufactured imports cheaper, damaging the local sales of our manufacturing industry. All of this increasingly creates a 'two speed' economy that disadvantages the majority of Australians who do not work or directly benefit from mining. The longer this goes on, the greater our dependency on mining.

Coal and climate

While many of the economic benefits of coal exports to Australia are grossly inflated, the damage done by coal is very real. Coal is the largest single source of greenhouse gases in the world and its consumption is increasing, not decreasing. Global demand for coal rose by 1 per cent each year from 1980 to 2000, then quadrupled to 4 per cent each year between 2000 and 2009.[16] In 2011 alone, world consumption of coal rose 5.4 per cent, powered by the expansion of China.[17] Australia's role is crucial in meeting this global demand. The Bureau of Resource and Energy Economics predicts coal exports from Australia of around 581 million tonnes a year by 2034–35. The coal industry wants even larger exports. This will mean that Australia will help produce more CO_2 emissions than Saudi Arabia currently does with its current massive exports of oil.[18] In doing so, we make the world's dirtiest energy easily available and encourage the developing world to industrialise in the worst possible way.

Many people still talk about the 'threat' of climate change, but in reality the changes are happening before our eyes. The most dramatic recent evidence is the rapid shrinkage of the once vast Arctic ice sheet. In the northern summer of 2012 the amount of Arctic ice shrank to the lowest level on record. Scientific forecasts had previously predicted that the Arctic could be free of ice in summer by around 2050. The current melting rate lends weight to the most pessimistic analyses which said that this may occur by the end of this decade.[19] Some climate scientists now see the decline in sea ice as linked to more extremes of heat and droughts in the Northern Hemisphere. If this is true, then we are already seeing the consequences for the world's poor, in wild swings in the price of staples such as corn and wheat.[20] The latest projections from reputable Australian bodies such as CSIRO express alarm: the world is emitting greenhouse gases faster than the worst-case scenarios previously predicted.[21] We are on track to a 4°–6°C warming by the end of this century; a catastrophic situation.

While climate change is occurring faster than scientists predicted, little is actually being done to actively lower global emissions, in spite of plans stretching back to the 1992 Rio summit. On the contrary, between 2000 and 2008 global greenhouse emissions rose 29 per cent.[22] This parallels the shift in the largest emission source from oil to coal. The world is bingeing on coal and unless coal is confronted there will be no solution to climate change. This was also the point made by former head of NASA's Goddard Institute for Space Studies, James Hansen, in 2009 when he said: 'coal is the single greatest threat to civilisation and all life on our planet'.[23] Ultimately, the world needs to phase out burning coal in order to avoid catastrophic climate change. It's one of a number of inconvenient truths.

Yet attempts to control and reduce the burning of coal have been feeble and ineffectual in most countries up to now. Put simply, one key reason is the power exerted by the global coal industry, increasingly referred to as Big Coal, on governments and political parties. You don't need to look very far to see the power of this group.

Big Coal stopped attempts to price carbon under the Howard government via the self-described 'greenhouse mafia' of industry lobbyists.[24] It attacked Labor's two attempts to price carbon. The first attempt was in 2009 when the Rudd government proposed an emissions trading scheme. This angered the coal industry because it was going to target the methane gas released by mining (called 'fugitive emissions'). Like the fossil fuel industry worldwide, the Australian coal industry fought this modest attempt to restrain greenhouse emissions by using its considerable wealth to put out misleading and alarmist information. Television ads proclaimed that more than a dozen coal mines would close and thousands of jobs would be lost. Two years later, after the defeat of the emissions trading scheme, the Labor government, supported by the Greens, introduced a carbon tax. Once again, the coal industry mounted an expensive television ad campaign highlighting predicted widespread mine closures and massive job losses, none of which came to pass.

Most action in Australia against climate change is aimed at decreasing our dependence on coal-fired electricity and promoting clean renewables such as wind power. This is valuable, but it is also necessary to look beyond Australia's shores. Climate change is a global problem. The consequences of a tonne of coal burned in New South Wales are no different for the global climate than a tonne of coal burned in Mumbai or Shanghai. That is why it is important to take a critical look at Australian coal exports, as well as the domestic burning of coal. There are two additional reasons why our coal exports are a problem. First, there is a significant degree of hypocrisy in Australia, where the use of green energy is encouraged at home while huge profits are made from exporting dirty coal. The greenhouse emissions from our coal exports nullify any planned emission reductions within Australia, many times over. In Australia and the United States, greenhouse gas emissions have decreased slightly in the last year or two, but this is a limited advantage in a world where the power plants of China and India are still gushing carbon dioxide into the atmosphere, encouraged in their dirty energy model by the easy availability of Australian coal.

Second, by placing long-term bets on coal expansion Australia is locking itself into a reliance on coal exports at a time when it is becoming clear that coal should be rapidly phased out. It makes no sense to allow the export boom in coal to structurally distort the Australian economy when it's possible that at some point an international agreement will emerge to discourage the use of coal. Such an agreement may be too late to prevent dangerous temperature rises, but when it occurs Australia's economic dependence on coal will make us far more vulnerable than we need to be.

But do Australia's coal exports really matter? The reality is they play a vital role in the global coal trade. This is because three countries (China, the United States and India) use nearly 5 of the 7 billion tonnes burned globally each year. Until recently they largely relied on their own mines. But now China and India are increasingly supplementing their own coal with imports. They join many other countries (Japan, South Korea, Taiwan) and the European Union (EU) which are already heavily reliant on imports. This makes Australia an important player because it produces one-third of the world's traded coal, including more than half of the coking coal exports used for steel-making. Global coal companies claim that coal consumption will rise to 11 billion tonnes a year in the next 15–20 years, with much of this coming from China and India. But it is unlikely that these countries will be able to provide this from their own resources, causing them to scour the globe for other sources of coal. This will make the growth in Australian coal increasingly pivotal. Without it, the world price of coal would be higher and coal's competitiveness would decline against cleaner alternatives.

This is most obviously true of electricity generation, but it also applies to the global steel industry, which is heavily dependent on Australian coking coal to meet its planned expansion. The coal industry has successfully persuaded many political leaders that you can't make steel without coking coal, but this is not true. Steel can be made from recycled scrap metal in electric arc furnaces and can run on renewable sources of electricity. Over a quarter of global steel production is done this way.[25] And there are alternatives to coal in other kinds of steel-making, most obviously using charcoal and natural gas. As the largest exporter of both iron ore and coking coal, Australia is playing a central role in ensuring that global steel production is, unnecessarily, as dirty as possible.

| | | |

For all of these reasons Australians must begin to think the unthinkable: if we are to prevent disastrous climate change we need a future that doesn't depend on coal. While this ultimately must be part of a global agreement, the Australian government can start restraining the export coal boom with a view to phasing out coal exports over the next couple of decades. Governments should refuse permission for new coal mines, prevent existing ones from being extended, and stop the subsidies to the coal industry. We need to phase out our own coal consumption and abandon any plans to build new rail lines and ports. At a global level, a positive step would be to price carbon high enough that it would discourage the trade in products requiring large amounts of fossil fuels. We need to view coal as the new tobacco or asbestos, a dangerous product whose use is strongly discouraged by the government and ultimately abandoned. Such suggestions are met with horror by coal corporations and their advocates, but given the disastrous likely effects of climate change, this is the only realistic position to adopt. Australia needs more of this kind of realism and less of the utopian belief that we can continue to mine and burn coal without catastrophic consequences.

1 THE REAL PRICE OF COAL

'Australia's response to climate change, both internationally and domestically, will be inextricably intertwined with the long-term future of the coal industry.'
ROSS GARNAUT, *The Garnaut Climate Change Review*

Australia is a land of coal. We are the world's second biggest exporter of coal and 75 per cent of our electricity comes from burning coal. Created over millions of years ago as lush forests and plants decayed and fossilised, coal was compressed into vast underground seams which lie beneath large parts of Australia. In Queensland, thick basins of black coal lie inland from the coast. In some places the seams are 40 metres thick, with mine sites forming a necklace that stretches for 800 kilometres. Further south, Victoria has some of the world's largest deposits of brown coal, centred around the Latrobe Valley, to the east of Melbourne. The valley's 65 billion tonnes of coal is often found just below the surface and up to 100 metres thick. In New South Wales, Sydney sits in the middle of a vast underground saucer of black coal, whose edges rise to the surface near Newcastle, Wollongong and Lithgow. Extensive mines honeycomb the land near each of these towns. The coal seams then travel north towards the Queensland border. In South Australia, black coal is mined at Leigh Creek to feed local power stations at Port Augusta. In Western Australia, coal is mined for electricity generation at Collie. Even in mostly hydro-powered Tasmania, coal is used for cement making.

Eighty years ago, the British writer George Orwell noted that 'our civilisation ... is founded on coal, more completely than one realises until one stops to think about it'.[1] Orwell could have been talking about Australia. In the early days of the Australian colonies, coal brought warmth on freezing nights and powered machines that replaced hard physical labour. In the nineteenth century, coal gas first lit the streets of Sydney and Melbourne. Train engines burned coal to haul wheat and wool to harbours where coal-burning ships took them half-way around the world. Later, coal began to boil water to spin the turbines to produce electricity. Coal used in steel-making was the foundation of much of the Australian manufacturing industry. Fifty years ago, a vast network of power stations was built to produce electricity for the refrigerators, the washing machines and toasters of the postwar consumer boom. In the twenty-first century coal still produces the electricity to operate our laptops, smart phones and flat-screen televisions. And now coal has become a major export industry. Little Coal has become Big Coal.

Coal has been an integral part of the European history of Australia from its earliest days when Lieutenant Shortland, hunting for escaped convicts in 1797, noticed some black rocks near a place he named Coal River, later called Newcastle, after the British coal port. Four years earlier, the French explorer Labillardiere noticed black horizontal veins near South Cape Bay in Van Diemen's Land (Tasmania).[2] The explorers Bass and Flinders, returning to Sydney Town from sailing around Van Diemen's Land in 1797, saw what looked like coal on the cliffs just south of Sydney. By 1804 coal was being dug up by convicts on the headlands of Coal River, which had become a prison within a prison, designed to punish convicts who had offended in Sydney Town. Some of the most active leaders of a rebellion by Irish convicts in 1804 were sent to dig coal there.[3]

Digging coal in those days was punishment indeed. The convict hewers of coal entered the pit by ladders and ropes and crawled through the metre-high tunnel until they reached the coalface. There they lay on their sides and hacked at the coal with picks by the flickering light of a candle. Other convicts bailed out the water that dripped through the mine and wheeled the coal to the bottom of the shaft, from where it was hauled up by a windlass to the surface.[4] For ten hours a day the convicts were expected to dig two and a half tons of coal.

Health problems became apparent as soon as mining began. In 1819 Commissioner Bigge was undertaking a major report on the settlement for the Colonial Office. The Government Surgeon in Newcastle, William Evans, told him, 'the foul air that is breathed [in the mine] produces spitting of blood and difficulty of breathing'.[5]

Elsewhere in the colony more coal was discovered. In 1824 it was found at Ipswich, west of Brisbane, and the following year at Cape Preston in Victoria. Coal was found south-east of Geraldton in WA in 1846 and, somewhat later, black coal was discovered at Leigh Creek in South Australia (1888).[6] But the area around Newcastle and the Hunter River was the main source of coal for most of the nineteenth century and well into the twentieth century. By the 1830s steam engines were being used to pump out water and several thousand tons of coal were mined each year. Coal was becoming popular in the expanding colony and by 1831 the first coal-burning steamships began to ply the coast of NSW.[7] Coal was displacing renewable energy forms, such as horses, human labour and the windmills that ground wheat. By the 1840s coal was providing energy for steam mills to grind grain and saw logs. In the fireplaces of homes it was gradually displacing wood, which was becoming increasingly expensive to cut and transport. The colony's most energy-intensive industries, lime burning and salt evaporation, operated at Newcastle close to the coal mines. In 1853, the *Sydney Morning Herald* installed the first steam-powered printing press in Australia with circulation soon rising from 1000 to 6000.[8] By the 1860s Australia experienced its first 'coal rush'. As the richness and vastness of the Hunter seams became more apparent and the price of coal rose, small mining companies scrambled to make money. In 1866 the first law restraining the use of coal was passed in the colony. It was aimed at stopping the dirty, sulphurous smoke caused by burning coal in Sydney.[9]

Coal exports also began early on. While much of the early convict-cut coal was sent to the colony at Sydney, some was sold overseas. When he reported the first coal discoveries, Governor Hunter was instructed to export coal to the Cape of Good Hope in southern Africa.[10] In 1824 an American ship took 250 tons of coal to Rio de Janeiro and in other ships coal was used as ballast and was exported to Bombay, Batavia (Djakarta) and Mauritius.

During the nineteenth century, sailing ships began to give way to coal-burning steamships. Australian coal, much of it from the rich seams of the Hunter Valley, supplied coaling stations in the Pacific and Southeast Asia. On land, from the 1860s railways spread from the coastal towns to the interior. Some early steam trains burned wood, but coal soon became the universal fuel. The railways developed a symbiotic relationship with the coal mines by hauling huge tonnages of newly dug coal and using coal-fired locomotives to do so. An early Commissioner of Railways spoke for many in 1875 when he gloried in the discoveries of the coal seams that stretched from Newcastle in the north to Wollongong in the south, and to the west of the colony at Lithgow:

> [The coalfields] are now computed ... to extend over an area of
> 15,400 square Miles and ... we may without exaggeration claim to be

in possession of the most valuable, most accessible and most extensive coal-fields in the Southern Hemisphere which must ultimately make New South Wales the richest of all the Australian Colonies.[11]

For a colony whose first mechanical energy sources were windmills and water mills, this promised much. Today, coal has even more enthusiasts and boosters. But there has always been a hidden price for digging and burning coal.

Paying the gas bill

Today one of the earliest uses of coal for energy has been largely forgotten. From the middle of the nineteenth century, coal was the key ingredient in a radical new technology that transformed the daily lives of Australians. This was coal gas (or town gas), which was produced when the black rock was intensely heated in enclosed iron chambers in the absence of oxygen. The roasting coal gave off a gas, which was then captured and purified. As it was being purified the gas gave off tar and ammonia and, later, even more toxic substances, such as naphthalene, benzene and toluene.

Gas-making first began on an industrial scale at a gasworks built on Darling Harbour in Sydney to supply street lighting through a network of pipes under the colonial town. The new gas technology evoked the kind of enthusiasm, hope and wonder, later heaped on succeeding technologies such as electricity and nuclear power. When gas lamps first lit the Sydney streets in 1841 the *Sydney Gazette* was lyrical about the 'wonderful achievement of scientific knowledge, assisted by mechanical ingenuity', and added that 'in the absence of our Almighty Maker's sun and moon, there is nothing like GAS'.[12] Indeed, gas was wonderful compared to the weak lamps in shops and streets that burned biofuels such as sperm whale oil or candles made from yellowing animal fat or tallow.

Gas quickly symbolised the aspiration to build a modern, progressive colony, but it also presented problems. The supply of gas was a natural monopoly and the lighting of streets was a public good. Later this meant that some gasworks were government-owned, but the earliest gas companies were private, profit-making ventures, which were often in conflict with the public authorities such as municipal councils. Gas companies relied on councils to allow them to dig up roads and bury their pipes. For their part, councils had no choice but to pay the price demanded by gas companies if they wanted to light their streets. Complaints from councils about high prices and the gas companies' monopolistic position were a feature of the times. At one point, the *Sun* newspaper argued that 'it is of course foolish and illogical to expect that a huge monopolistic concern like the Australian Gas Light Co. has any wish other than to fill the pockets of its shareholders. To expect [otherwise] would require the faith of a man who looks for mercy in a shark ...'[13] Shortly after this, the government moved to control both the price of gas and the profits of the gas company.

The steady march of coal gas soon reached all capital cities. In 1856 the Melbourne Gas and Coke Company first illuminated the city's streets. The following year gas came to Hobart, where, as well as illuminating street lights, it enabled a butcher to give demonstrations in his window of how to cook meat using a gas flame. In 1863 Adelaide began producing gas from a gasworks at Brompton and in Brisbane the first gas was produced in 1865, initially using coal imported from Britain, rather than developing local reserves at Ipswich. In 1885 Perth was the last capital city to start manufacturing coal gas.[14]

Gas changed the look and feel of Australian society. Walking on the streets after dark lost some of its fears after the small army of lamplighters moved through them, returning at dawn to extinguish the small burning flames.

By 1890 the use of coal gas was booming in places like Sydney where each night 9000 public gas lamps burned and gas was piped to 27000 homes. Australian Gas Light (AGL) was busily installing new retorts to cook hundreds of tons of coal to meet an annual rise of 20 per cent in demand.[15] In succeeding decades other gasworks sprang up on the harbour shores at Manly, Waverton and on the Parramatta River at Mortlake. During the early twentieth century, gas for illumination was unable to fight off the challenge from another coal-based rival, electricity, but it remained a source of energy for cooking and heating water. The other rival to coal gas was 'natural gas', derived from petroleum deposits, and from the 1960s it began to replace coal gas in kitchens and factories.

Today coal is no longer used in Australia to produce gas in this way, but coal gas is still very much with us. Coal gas left a toxic heritage. Until recently, in Sydney, Melbourne, Brisbane, Adelaide, Perth and dozens of rural towns, the sites of old gasworks oozed dangerous chemicals and tar. This was the real price of coal, unpaid for many decades until it finally caught up with us.

The price of the gas bill has been hundreds of millions of dollars, paid by state governments to clean up these poisoned sites. According to a 2005 study of rehabilitating old gasworks, such sites are 'invariably contaminated' by such things as heavy metals and cyanide.[16] These wastes form a 'black odorous ooze and iron cyanide complexes ... recognisable by their intense Prussian blue colour'. Most gasworks dumped their tar in deep pits and these contaminate local ground water. This was a problem at the old East Perth gasworks site. Cleaned up in 1994 for $17.5 million, it was discovered the site was still spreading toxic chemicals to fish and mussels in the Swan River when tests were carried out in 2011.[17]

Nearly all Australian gasworks were built on rivers and harbours so that coal could be shipped in easily. Today the old gasworks sites have become prime waterfront real estate, such as Melbourne's Docklands. When the Victorian authorities first proposed developing waterfront land, mouthing the usual real estate boosterism, they were soon confronted by the land's toxicity. One local journalist at the time commented that sites like Docklands had everything. They had proximity to the beach and rising property values – as well as cyanide, lead, coal tar residue, naphthalene and much more.[18] All courtesy of the gasworks that operated from the 1880s to 1955. The cost of the gasworks clean-up was at least $46 million, paid for by Victorian tax-payers.[19] On the Brisbane River, the luxury homes of Newstead Riverpark now sit on the old gasworks site, reckoned to be one of the biggest remediations ever undertaken in Australia involving the removal of 1 million tonnes of soil.[20]

In Sydney today the city's proudest development is Barangaroo, a 22-hectare site on Darling Harbour named after one of the early Indigenous inhabitants. Old concrete wharves are being converted to parks and buildings. But part of Barangaroo sits on top of the site of the gasworks which first lit Sydney streets in 1841. In 2004 residents of a nearby apartment block complained of strange odours and headaches. Tests later discovered the presence of cyanide, toluene, benzene, copper and lead close to the underground car park. The area for the flagship project was once known as the Hungry Mile because wharfies looked for work there during the Great Depression. Later it was dubbed the Toxic Mile and the clean-up bill estimated at $100 million.[21]

Gas in the mines

But while the toxic price of coal-fed gasworks only became apparent 150 years later, another price was obvious right from the start. It affected the lives and health of the coal miners and the communities that often clustered around mines. For all of the nineteenth century and the early part of the twentieth century, coal mining was a primitive process. A coal seam was identified and a shaft dug. Wooden props held up the roof, while ponies or pit boys wheeled the heavy skips along on crude rails to the entrance where it was sorted and sent on trains or ships.

Coal mining in Australia was more dangerous than in Britain and attempts to make it safer met with the determined opposition of the 'coal masters', the precursors of today's fossil fuel giants. While roof collapses killed many miners, the biggest killer was coal seam gas or methane. When ignited, the gas usually caused a gigantic explosion and sucked oxygen from the mine. Reporting on the 1887 mine explosion at Bulli on the NSW south coast, the *Sydney Morning Herald* said:

> In the fiery blast that belched forth from its unknown seat in the bowels of the mountain every object was torn and twirled and wrenched and smitten ... of the poor remnants of humanity that were shrivelled and battered by its cruel might ... imagine a little moth imprisoned in the powder charge of a huge cannon ...[22]

A Royal Commission following the Bulli disaster found that it killed 81 miners, widowed 37 women and left 120 children fatherless. The year before 29 miners had been killed in NSW and the year after the Bulli disaster 41 more miners were killed.[23]

Attempts to control the carnage in the coal mines were loudly resisted by the mine owners. The 1894 Coal Mines Regulation Bill insisted on a minimum of 150 cubic feet of air for each miner and limited the working day to eight hours. Other provisions called for the fair weighing of coal (which determined the coal miners' pay) and safety procedures, particularly about the use of miners' lamps. The coal masters campaigned against this attempt to reform the law and protect the miners. One MP with coal interests, William McMillan (dubbed McMillion by the *Bulletin*), described the Bill as an attack on 'individual liberty'.[24] The Bill was gutted by the Free Trade MPs, but an election soon after gave the government a mandate for the

Bill. It appointed a further Royal Commission which supported the Bill.

Even when passed in 1896, the clause for an eight-hour day was removed and 'adequate ventilation' was defined at 100, not 150 cubic feet of air. A number of mine owners immediately challenged the Bill in the courts, while others simply broke the law and refused to implement fair weighing of coal. A few years later the *Miners Accident Relief Act* laid the basis for later laws on compensation for workplace injuries. Just six years after the first simple reforms to make the mines safer, another terrible explosion occurred on the NSW south coast. In July 1902 gas ignited within the Mount Kembla mine and 96 miners were killed. A later Royal Commission heard evidence about poor testing for gas and the mine's inadequate ventilation.

For such miners, the price of coal was their lives. And gas explosions have not stopped. In 1972, 17 miners were killed near Ipswich when a mine feeding the Swanbank power station exploded. In 1975, 13 miners died when the Kianga mine in Central Queensland exploded. In 1979, 14 miners were killed at Appin, south-west of Sydney and in 1986, at Moura in Central Queensland, 12 miners died with another 11 miners killed there in 1994.

With battles over health and safety, the history of the coal fields has been one of bitter conflict between the mine owners and the coal miners. The historian, Robin Gollan, studied the early coal mining industry and declared, 'More than in any other industry, employers and employees have faced each other as enemies, and fought each other with all the bitterness, the high hopes, the despair and the suffering of civil war.'[25] In 1940, Judge Drake-Brockman of the Arbitration Court described the coalfields as the scene of: 'an unbridled and unregulated struggle between employers and employees without restraint and actuated only by the rules of the jungle'.[26] Today the owners of coal mines are often remote global corporations, but they are every bit as ruthless as their ancestors, bullying any government that moves to price carbon, to tax their profits, to stipulate working conditions or regulate their actions.

Dust and ashes

For the men who dug the coal below ground, the price of coal went far beyond conflicts over gas ventilation or inadequate pay. While some miners paid with their lives, far more paid with their health. Even without roof collapses or gas explosions, 'normal' coal mining extracted a debilitating price. Gollan bluntly described the situation: 'From the earliest times mine-owners had shown a complete indifference to the health and safety of their workers'.[27] No better illustration of this was the fight put up by mine owners against measures to prevent dust disease or black lung, as it was known in the United States, because of the colour of the miners' lungs when examined at post mortem.

An insidious evil, the effect of coal dust on the lungs only began to appear after miners had spent many years at work. When coal was dug underground a fine dust was thrown up and in the confined spaces of a mine those working at the coalface inhaled it constantly. The result was a scarring of the lungs as the microscopic coal dust permeated the living tissue. Through the 1920s and 1930s, as the number of gasping, dying coal miners grew, the mine owners fought the decisions of workers' compensation courts, which often ruled in favour of these desperate workers. Medical opinion in Australia was divided on whether coal dust was dangerous to miners. Some doctors, often those retained by mine owners' insurance schemes, claimed that coal dust was 'inert', thus creating a legal loophole which meant no compensation would be paid.[28]

A 1939 Royal Commission into health and safety in coal mining examined the issue of rock collapses, methane gas and other dangers in mines. But the most contentious issue was coal dust and its effects on miners. The Inquiry heard evidence that in some mines the dust was so thick that 'men at work on the face cannot see his mate who is situated only a few yards away'.[29] The Royal Commission acknowledged the lack of a consensus of medical opinion on the dangers of coal dust, but pointed out that in naturally wet mines the incidence of lung disease was much smaller. It recommended the use of pressurised water at coal faces to reduce the 'dust hazard', but such changes were barely implemented as Australia's coal mines soon went into high production for the war effort.[30] After the Second World War, it became increasingly obvious that the damage to miners' lungs was worsening. The *Sunday Telegraph*, on 7 July 1946, reported that:

> the menace of dust on the lungs has caused near panic among miners on the South Coast of New South Wales. Three of them died last week. None of them know who will be next ... George Bryson, 41, of Douglas St, Wollongong, coalminer, has a wife and two children .... At first sight you would take him for a man in the prime of his life with many years of hard work ahead of him. But Bryson has dusted lungs .... He's one of the walking dead of the South Coast. There are about 1,000 others like him on the southern coalfields.[31]

A second Royal Commission in 1946 inquired into the issue and commissioned a group of health experts who strengthened the case that coal dust was deadly. The mine owners, however, complained about the 'heavy increase in the cost of workers' compensation', which they said was 'largely resulting from the dust hysteria'.[32] Prime Minister Ben Chifley's Labor government rejected this and began a series of reforms and regulations that led to the virtual elimination of lung disease from airborne coal dust suffered by underground miners in Australia. By contrast, in the United States, where coal companies successfully prevented measures to dampen down coal dust until recently, retired miners still die in large numbers. In the ten years to 2009 about 10000 people (mostly retired miners) died in the United

States from the effects of coal dust, according to the National Institute of Occupational Safety and Health.[33] Just as the coal companies denied for many years the medical consequences of coal dust, so today they still deny the findings of climate science.

The price of electricity

In 1882 a wondrous new invention was exhibited in Sydney Town Hall. Built by the American inventor, Thomas Edison, it consisted of a glass bulb containing a carbonised filament, which, observers noted, had the brightness of eight candles when an electric current was passed through it. 'The light provides no sensation of heat and the electricity can be run into a block of houses in the same manner as gas,' said one report.[34] Shortly afterwards, the main post office in Sydney replaced its gas lights with electric ones, nevertheless powering them with a gas generator.

Beginning as a novelty, during the early twentieth century electricity became widely seen as a symbol of social and political progress. Gas was increasingly seen as dingy and dangerous. Electricity was clean, safe, bright, instant and wrapped in an aura of a glowing technological future. The new technology of electricity generation also appealed to country and city municipalities because they could own and control this new energy system. Public ownership meant they could achieve independence from the private, for-profit gas companies.[35] The coming of electricity also signalled a move from the bio-fuelled Age of the Horse. While trains carried heavy loads for long distances, for short journeys the early cabs, buses and even trams were horse-drawn. By the turn of the twentieth century, electric trams began to spread their rails into the corners of many Australian capital cities. Until the end of the First World War most of the electricity produced in Sydney was used to power its electric tram system.

The continued reliance of trains on coal meant that the soot and dirt of trains were scattered over houses and shops built near the railway lines. Soon, suburban trains were electrified, reinforcing the perception of electricity as the clean fuel of the future. Electricity also delivered fun and entertainment. In the crowded cities it enabled spectacular displays of public lighting on an unimagined scale. Shops and cinemas were transformed. Dazzling electric lights replaced the soft, dim gas lamps. Theatres devoted to 'talking pictures' relied on electricity to run their projectors and woo their customers. Ballrooms sparkled with incandescent light and the dance bands and singers' voices were amplified by electricity. The giant grotesque face on Sydney's Luna Park was illuminated by hundreds of glowing lights.[36] For all this, electricity came from dirty coal-burning power stations. The exception was Tasmania where hydro electricity was developed after first being produced in 1895. Remarkably, hydro electricity was also created from 1893 to 1951 at Thargomindah in western Queensland by harnessing an artesian bore that also provided the town's water supply.[37]

The Second World War was a turning point for Australia. The military threat from Japan meant that a nation known for its wool and wheat moved decisively to large-scale industrialisation, symbolised by the rapid expansion of steelworks in Newcastle and Port Kembla. Coal-fired steelworks could provide security via ships, tanks and the weapons of war. Powering the change was intensive electrification spreading beyond major Australian cities to regions and the bush. New power stations tended to be built on top of coal fields rather than near cities. Through the 1950s and 1960s the hunger for electricity grew rapidly and consumption doubled every decade. This growth fed the cornucopia of things powered by the magic of electricity: washing machines, clothing irons, vacuum cleaners, radios, refrigerators, power tools and heaters. Today the flood continues with dishwashers, computers, televisions, bread makers, food blenders and other electrical gadgets. With smoky power stations banished from major cities, electricity has entrenched its aura of a clean and reliable form of energy.

But electricity produced by coal combustion still comes with a long-term price. Today, medical researchers are finding increasing evidence that human health is damaged by the air pollution produced by coal-burning plants. In 2009, the US-based Physicians for Social Responsibility compiled all the peer-reviewed medical research on the health effects of air pollution caused by coal, and analysed the results. They published their findings in *Coal's Assault on Human Health*. The respected physicians group (which won a Nobel prize in 1985) point out that burning coal 'releases a combination of toxic chemicals into the environment ... [such as] sulphur dioxide, particulate matter (PM), nitrogen oxides, mercury and dozens of other substances known to be hazardous to human health'.[38] These pollutants damage the lungs, heart and nervous systems of some people and contribute to some of the leading causes of death in the United States: heart disease, cancer, stroke and chronic respiratory diseases.

These adverse effects should come as no surprise. In 1952, London experienced a 'killer fog' which lasted four days and sent death rates and hospital admissions soaring. Almost 12 000 deaths were attributed to this environmental disaster, caused, in part, by burning coal for domestic use.[39] More recently, the link between burning coal and health was made clear in Dublin, Ireland, in the 1990s. In the 1980s, the cost of fuel oil rose and many Dubliners switched from oil to coal to heat their homes and provide hot water. But the ensuing air pollution was associated with an increase in in-hospital deaths due to respiratory diseases. According to the physicians group:

> This led the Irish government to ban the marketing, sale and distribution of bituminous coal on September 1, 1990. In the year that followed, black smoke concentrations declined by 70 per cent, respiratory

deaths fell by 15.5 per cent and cardiovascular deaths fell by 10.3 per cent. Approximately 450 lives were calculated to be saved that year by this measure.[40]

Perhaps the most alarming effect of pollution composed of material produced by burning coal is on children. Burning coal produces pollutants like nitrous oxide and particles as small as 2.5 microns wide (a human hair is 18 microns), and both of these adversely affect lung development, according to a study of 1700 children in California. Air pollution triggers attacks of asthma, and children, partly because of such things as the time they spend outdoors and their immature immune systems, are highly susceptible to pollutants such as those emitted by coal-fired plants.[41] For adults, exposure to nitrous oxide and tiny particulate matter is correlated with the development of lung cancer. Recent research suggests these same pollutants are associated with heart attacks and related problems.[42] Children's nervous systems can also be affected. Coal contains tiny amounts of mercury and when burned, mercury enters the atmosphere and returns to earth in rain. As it travels up the food chain, mercury increases in concentration and can reach high levels in large fish. Humans are exposed to coal-related mercury largely by eating fish. A large 1999–2000 study, reported by the Physicians for Social Responsibility, showed that 15.7 per cent of US women of childbearing age were found to have blood mercury levels that would cause them to give birth to children with mercury levels exceeding the recommended maximum. Researchers have estimated that between 317 000 and 631 000 children are born in the United States each year with blood mercury levels high enough to cause lifelong loss of intelligence.[43]

In the NSW Hunter Valley coal is both mined and burned in power stations. The air breathed by residents is sometimes a toxic cocktail of dust thrown up by explosions (used to loosen coal at open-cut mines) and the gases emitted from the two coal-fired power stations on Lake Liddell. In 2012 a group from Sydney University reviewed 50 studies of the health and social impacts of coal mining and coal combustion. Their report, *Health and Social Harms of Mining in Local Communities: Spotlight on the Hunter region*, highlighted health effects such as excess deaths and increased rates of cancer, heart, lung and kidney disease and birth defects.[44] In 2008 a study showed that 113 tonnes of toxic metals (such as antimony, arsenic, cadmium, chromium, lead and mercury) were emitted into the air of the Upper Hunter, along with over 13 200 tonnes of sulphur dioxide and 62 000 tonnes of oxides of nitrogen.[45] Melissa Mattson, a mother who lives in the Hunter town of Singleton, described how her three children all suffer from skin rashes and allergies. 'We sent [them] on a family holiday to Darwin in June–July last year and within three to four days every single rash cleared up. When we came back it only took two to three days again for them to come back.'[46] A doctor in Singleton, Craig Barry, noticed a similar thing when his children went on holidays. When they returned they had to use puffers and nasal sprays for their breathing difficulties.[47] While they have a genetic base, asthma, allergies and eczema are aggravated by local conditions.

In 2009, the NSW state government agreed to install 14 monitors to check dust levels, but only three of them are designed to measure the tiniest (2.5 micron) and most dangerous particles. It is worse in Queensland, where only two of the 29 dust monitors installed by the Queensland government are in mining districts, the rest being in major towns. The one monitor in a mining area operates in Moranbah in the Bowen Basin, but its results are not publicly released.[48] In Victoria's Latrobe Valley, where brown coal is burned, a study found there 'were significant associations between airborne particles, nitrogen dioxide, and respiratory morbidity', and that these pollutants were associated with hospital admissions for diseases like bronchitis and emphysema.[49]

Research in Australia on the effects of coal on health is not extensive, but there is little reason to think that the results differ from the United States. A group of Australian doctors reviewed the evidence here and overseas, and concluded that 'there is overwhelming evidence that coal mining and the burning of coal is harmful to physical and environmental health, and can have a significant impact on local communities'.[50] Australia is an advanced country capable of phasing out the use of coal and replacing it with non-polluting technology, say the doctors.

Today's export coal rush

In the last ten years coal mining in Australia has been transformed. In a relatively short period we have become the world's second biggest exporter of black coal and the world's fourth largest producer overall, behind China, the United States and India. In 1991 Australia exported 120 million tonnes of coal. By 2012 it had grown to 316 million tonnes.[51] By 2025, big mining companies hope to export between 520 and 689 million tonnes to the rest of the world, according to one government estimate.[52] Today's 'coal rush' is a central part of what a former deputy governor of the Reserve Bank has called a 'wild west resource stampede'.[53]

The origins of the current export boom lie in the 1960s with the growth of the Japanese steel industry fed by Australian coking coal, particularly from Queensland. In 1960, prior to the Japanese steel boom, we exported a tiny 1.9 million tonnes.[54] The steel boom saw the arrival of overseas companies like Utah, Peabody, CRA and Mitsui, joining BHP and Thiess. In the 1970s the profits repatriated by Utah alone were so huge that they began to affect the value of the Australian dollar and the level of foreign reserves. In part, the boom was fuelled by the 'oil shocks' of the 1970s, which saw Asian electricity producers switch from oil to coal-fired power, offering a growing market for Australian thermal coal. This period also saw steel mills in Taiwan and Korea add to the demand for Australian coking coal. The growing coal boom saw the spread of open-cut mining, a method in which millions of tonnes of 'overburden' – soil, trees and rock – are ripped off the surface to expose the underlying seams of coal. The coal is then scooped out by draglines, giant bucket shovels that move across the coal seam, often 24 hours a day, seven days a week.

While coal mining expanded throughout the 1980s and 1990s, the export price for Australian coal declined as the price of oil dropped and some energy users switched back to oil. In the 1990s coal's future looked bleak and a number of oil companies, such as Exxon, sold their coal mines to the 'Big Four' of Big Coal in Australia: BHP, Rio Tinto, Xstrata and Anglo American. By 2006 the Big Four dominated Australia's coal-mining industry and produced 74 per cent of total saleable production, with each commanding between 17 and 19 per cent.[55] Such concentration is 'extraordinarily high for any industry,' noted one US analyst.[56]

The Big Four bet big on coal and won. From 2003 the world price of coal began to rise rapidly, boosted by Chinese demand. From around US$25 per tonne in 2003 thermal coal quadrupled in price to well over $100 per tonne in 2012.[57] In this boom, the global financial crisis of 2008 was merely a hiccup. A government survey of the mining and energy boom noted that in the first half of 2011 ten new coal projects reached an advanced stage, with 14 more in the pipeline.[58] By the end of 2011 there were six port expansions and four rail expansions under way – all to service the coal rush in Queensland and the Hunter Valley. In the mining industry overall, new spending in 2010–11 was $55 billion, a jump of 53 per cent on the previous year. Spending in 2011–12 was forecast to leap again to a mind-boggling $73.7 billion.[59] All of this is transforming Australia, both economically and physically. But in some areas of rural Australia mining is having a devastating effect (see Chapter 2).

Overall, the mining boom, with coal and iron ore at its heart, is beginning to reshape the Australian economy. The boom in coal mining is not merely another useful addition to other sectors of the economy, rather, it is in competition with them. The headlong expansion of the mining industry is sucking up capital and construction capacity, as well as labour. Money for roads and rail, for ports and bridges, is being used to service the coal industry rather than supporting other industries. One small but telling example: farmers on the Darling Downs and elsewhere in Queensland now have to pay more to ship their grain by road because mining companies monopolise the rail network.[60] One farming body estimates that 70 per cent of all grain in Queensland has now been forced onto the roads.

On an even more significant level, the mining industry competes with agriculture and other export industries such as tourism and education, which are also more labour intensive.

But the coal industry and its partners, such as coal-based electricity generation, are promoting another transformation in contemporary Australia, which will have an even greater effect. When coal is burned in the world's numerous power stations, the atoms of carbon join those of oxygen to form carbon dioxide. Climate sceptics triumphantly describe CO_2 as 'plant food', and indeed it is. Without carbon dioxide, life on planet Earth would be impossible. Balance is all: the planet Mars has only a thin layer of carbon dioxide and its average temperature is around minus 50°C. On the other hand, Venus has a dense blanket of carbon dioxide and it averages 450°C.

Paying the final price

The burning of fossil fuels like coal and oil has already warmed Australia and will continue to do so. The average annual temperature in Australia has risen by just under one degree (0.9 °C) since 1910 and, in less than 20 years' time (2030) it will rise another degree above 1990 levels.⁶¹ What happens after that is largely dependent on what we do now to reduce the amount of coal and oil that we burn and export.

We can already detect the consequences of climate change in our daily lives. Well-regarded studies by CSIRO and the Bureau of Meteorology show that the current number of hot days and warm nights per year has increased since 1955 and heatwaves are now more common.⁶² One such heatwave in February 2004 meant that Sydney had 10 successive nights over 22°C (the previous record was six). Adelaide suffered more, having 17 successive days over 30°C (the previous record was 14 days). In 2008 Adelaide had 15 consecutive days of 35°C or above, surpassing the previous record of eight days. Such heatwaves will intensify, according to the latest scientific research. With no mitigation of emissions, the number of days with temperatures over 35°C in Perth would rise from 27 currently to 72 days in 2100. In Adelaide it would rise from 17 to 44. In Darwin, by 2100, fewer than eight weeks of the year would consist of days below 35 degrees.⁶³ Among other things, this will mean that bushfire seasons will start earlier, end later and be more intense.

While higher temperatures can lead to more evaporation and more rainfall in some areas, CSIRO and the Bureau of Meteorology have concluded that recent droughts in Australia are linked to, or exacerbated by, global warming. For example, scientists have found that rainfall in south-eastern Australia has declined. From 1997 to 2007 only one year had rainfall above the 1961–90 annual average.⁶⁴ This has meant less water in streams feeding the dams of major cities. In Sydney this has meant streamflow is 40 per cent of the long-term average, while in Melbourne it is 65 per cent. The worst drop has been in the streams feeding Perth's water supply, which in recent years has caused their streamflow to be only 25 per cent of the long-term average.⁶⁵ In January and February 2013 Australia had a taste of the future. We experienced our hottest summer ever, with the hottest day ever recorded occurring on 7 January. Many other records, including those for sea temperatures, were also broken, report climate researchers.[66]

Overall, *The Garnaut Climate Change Review* says that without an effective reduction in carbon emissions the impacts of climate change on Australia 'are likely to be severe'. By mid-century we would face 'major declines in agricultural production' with irrigated agriculture in the Murray-Darling Basin losing half its annual output.[67] At the same time, with no mitigation of greenhouse gases we would see 'effective destruction of the Great Barrier Reef'. By 2100, with no mitigation, irrigated agriculture in the Murray-Darling Basin 'is likely to have ended' and 'depopulation will be underway'. If the world does little to reduce the burning of fossil fuel, other effects could include the devastation of the wheat crop, big losses in the tourism industry, coastal flooding and a rise in extinction rates. In Asia and the Pacific, crises in food production, sea level rises and the growth of 'climate refugees' would destabilise the region and have serious consequences for Australia's national security.[68]

Today Australia is feeding the world's lust for coal. The carbon produced by the burning of Australian coal is beginning to melt glaciers, boost sea levels and play havoc with the climate of the Earth. Australia, along with the rest of the world, is poised to pay the final price of coal.

2 THE STRIP-MINING OF AUSTRALIA

'A foreign company is here, quite frankly, to make money at our expense. Any benefits to us are incidental. Its policies and decisions are formulated in London or New York or Geneva ... its directors often would not know, and most often would not care what harm they do to us, or our environment, provided they continue to make profits ...'
EDWARD ST JOHN, QC, former Liberal MP [1]

Among the gum trees a sandstone cave sits high on a hill above the broad valley of the Hunter River in NSW. Inside the cave, the figure of a man – or rather a god – is painted in red ochre. Around him are white stencils of human hands, a stone axe and a boomerang. It's an ancient vision of the Indigenous spirit god Baiame, who came to Earth and shaped the rivers, the hills and the forests. However, today, if you stand in this cave and look north across the Hunter Valley you see another vision.

Soft grey mist suffuses the horizon, but it doesn't obscure the distant towers of power stations, the serried grey ridges and undulating foothills in the distance. When you take a closer look, the ridges and foothills are not what they appear. They are giant heaps of overburden, the rock and soil scooped up by the huge draglines that excavate the rock and soil as they expose the coal beneath. The grey mist is not natural. It is the dust kicked up by dozens of draglines and hundreds of giant trucks, aided by hundreds of tonnes of explosives that shatter the rock. All this feeds the conveyor belts and hoppers which send the coal on its journey to Japan, South Korea and Taiwan. The scars of the valley's ugly mines are usually screened by a ribbon of roadside trees, so the devastation is best seen from above, on satellite photos from Google. From above they resemble grey tumours and abscesses scattered over a landscape where the dairies, horse studs and vineyards are left to fight a losing battle.

In the Hunter Valley there is an urgency, amounting to desperation, to dig up the coal and feed it into the low-slung, fat-bellied ships that wait at Newcastle's port (now the largest coal port in the world). Much more is on the way. Rail lines have been doubled in number, a fourth coal loader is proposed, local objections brushed aside. According to one report the plan is to have a train disgorging its coal at the port every 15 minutes.[2] Newcastle and its Hunter hinterland are in the midst of a growing coal rush. In the last ten years only a couple of proposed mines have ever been rejected by the state government. The Hunter region holds most of the 60 coal mines in the state of NSW and produces over 70 per cent of its coal. Coal production has almost tripled since the early 1980s and will probably double in the coming decade. Five hundred square kilometres of the region are directly affected by mining. One billion dollars a year is creamed off by the NSW government in mining royalties Many more billions of dollars are pocketed each year by the giant, often foreign, coal-mining companies.

This vision of the Hunter Valley is one that Australians who live far from the region should understand and be wary of. It is a vision of a future that is coming to dozens of rural towns and valleys in NSW and other states. Thanks to accidents of geology and the rising world demand for coal, vast swathes of rural Australia are on the road to becoming like the Hunter Valley. In some areas, local farmers and country people are fighting the early assaults of Big Coal. In other places, such as Central Queensland, coal mining has existed for years, but it is now furiously expanding.

If we want to examine the real cost of coal, both in human and economic terms, the best place to start is the Hunter Valley, where it has been mined for more than 100 years and has exploded in the last decade.

To call it a valley perhaps gives the wrong impression. For the most part, the Hunter Valley is a broad, flat floodplain formed over thousands of years. This plain has supported dairies, vineyards, horse breeding and coal mining over the last century. Small rural villages and towns often grew up around the underground coal mines. The mines were often originally worked by Welsh and Scottish coal miners, who named the villages in memory of their homeland: Abermain, Standford Merthyr, Greta, Neath. As recently as the 1990s, coal mining in the Hunter was a modest affair with miners being laid off when a downturn emerged or the demand for coal dipped.

However, the old mines have given way to something new in recent years. The mining companies soon re-employed the miners they had sacked a few years earlier. In 2005 a local journalist, Greg Ray, described the looming transformation, which has since come to pass:

> The super pits of the 21st century are truly huge. They shift huge amounts of earth and they stand out in ways that the mines of the past never did. Whole forests are moved, whole communities are bought out, streams diverted and very big holes made.[3]

Since 2005 the coal rush has only intensified. Where once a mine produced 1 or 2 million tonnes per annum, today Xstrata's Mangoola (Anvil Hill) mine produces over 10 million tonnes per year, while BHP Billiton's Mount Arthur mine is well on its way to producing up to 30 million tonnes per year.

A boom for some is a bust for others. A farmer for over 30 years, Wendy Bowman has seen the Hunter Valley transformed. Twice she has had to sell properties where she farmed because of the encroachment of coal mining. Now she lives near the village of Camberwell, itself being slowly surrounded by mines.

Her first inkling that coal mining would come to dominate her life and her business came in 1988 while she was running a dairy farm at Ashton, a property near Glennies Creek.[4] She noticed the cows were sometimes refusing to eat new green grass, and when they did, they ate it in what appeared to be circular patterns. After much study a farming expert discovered the answer. Dust from a nearby mine was being held in the furry tips of the grass (actually a form of barley). Eddies of wind shook the dust off some of the new shoots and, sensing this, the cows ate only the dust-free grass. Later her milk was rejected by the dairy company because it contained dust blown across the valley from open-cut mines.

Then a neighbour told her of other unusual problems. He suddenly found it impossible to grow lucerne along a creek where he had grown it for years. Some distance up the creek they found its flowing water disappearing into a crack in the stream bed. The water re-emerged further down the creek, but when it did it was more salty and would not sustain the growth of lucerne. The reason for this soon became clear. A mine had been dug underneath the creek and had cracked the stream bed.

While she was running her dairy herd at Ashton, Wendy Bowman had been living on an old homestead at Rix's Creek, across the valley. The experience of the relentless pressure of mining led her to form Minewatch, a community group with the aim of defending the valley from unrestrained mining. But she soon discovered that she would have to defend herself because she would soon have a coal mine as a neighbour. Her activities with Minewatch also meant that other residents came to her with unusual problems. A neighbour found that his Black Angus cattle were not black at all, but rather a mottled grey with patchy hair and suffering from diarrhoea. A vet took blood samples and concluded that the soil contained no copper, a trace element needed for healthy animals and for colour in their hides. Normal soils contain tiny amounts of copper and many other trace elements, but this was not normal soil. It was 'rehab', the name given to land left after mining and 'rehabilitated' by the departing mining companies.

Meanwhile, at her home on Rix's Creek, Wendy Bowman found the sound of blasting and the dust whipped up by the new mine becoming intolerable. She finally sold the old homestead to the mining company in 2005, which promptly demolished the house and her beloved gardens. Today in Camberwell the noose is tightening once more, as the inexorable march of mining surrounds her a second time. Soon a mine will be within 500 metres of the village on two sides and 1 kilometre on the third.

Sacred sites

In 2010 the NSW state government handed over the Camberwell village common to Ashton Mining (owned by the Chinese-backed Yancoal). The following year it approved a 1.7 kilometre diversion of Bowman's Creek to allow Ashton to mine another 5.3 million tonnes of coal. But this underground mining will cause subsidence and damage a series of signficant Aboriginal sites. An archaeological survey map from the Department of Environment shows them dotted along Bowman's Creek.[5] Some are ancient axe-grinding grooves, stone fish traps, while others contain small artefacts. A local Indigenous man, Scott Franks, is angry. 'Ashton says it's OK because they'll "remediate the site". How do you remediate the destruction of a sacred site?'

There is another significant site in the Hunter Valley. Wambo Homestead, built in the 1840s, is a grand Victorian Regency farmhouse of colonial Australia, a rarity which still survives and is thus listed on the NSW heritage register. But it sits on 20 million tonnes of high-quality coal worth $2 billion, which the company Wambo Coal wants to mine. Owned by the global giant, Peabody, and Japan's Sumiseki, Wambo Coal applied to de-list the historic homestead in 2010. The mining giants denied its heritage value and complained that they would suffer 'undue financial hardship' unless the homestead was de-listed.[6] The mining giant has offered money for ways to remember the colonial homestead after its demolition, suggesting ideas such as 'virtual visits' online.

The attempt at de-listing, according to local retired history teacher, Carol Russell, 'is a test case of whether we are to be sacrificed to God Coal'. For the moment, Wambo Coal has backed off and the heritage listing remains on the old homestead, which is nevertheless likely to be surrounded by mining, including underground mining, nearby.[7]

The rise of Big Coal in the Hunter is not simply a matter of threatening the natural or human heritage of the area. Coal mining displaces and damages other businesses, such as agriculture and tourism. One of the strong opponents of further coal mining and coal seam gas exploration is the thoroughbred horse industry. The Hunter Thoroughbred Breeders' Association represents a multi-billion dollar industry, second only to Kentucky in the United States. Mining now encroaches within hundreds of metres of some horse-breeding boundaries and the industry says it is now at a 'tipping point'.[8] The horse breeders fought the NSW Labor government's approval of the massive expansion of mining. Now they are angry that the new Liberal–National government has broken its promises to restrain the coal industry. Its new policy offers a 2-kilometre buffer zone around mines, which is meant to limit the vast amounts of wind-blown dust. But such a buffer is meaningless, according to the Breeders' Association vice-president Andy Wiles, '…we already get dust from mines that are eight kilometres away,' he says.[9]

The horse breeders were joined by the winemakers, representing another world-famous industry of the Hunter Valley. The owner of the long-established Tyrrell's Wines, Bruce Tyrrell warned,

> During the 1980s the mining industry sweet-talked the Government and wine industry into opening up the Upper Hunter region using the promise of 'co-existence'. Look what happened. The Upper Hunter has been lost to wine and tourism completely. It has been consumed by coal mining and become an industrial – not recreational – playground.[10]

Another winemaker who has been stung by Big Coal is John Cruickshank, who was forced to sell his Callatoota Estate vineyard when the NSW Labor government approved the Anvil Hill mine in 2007. Later, Centennial Coal bought land from him for $2 million. 'It wasn't enough money, but they had us over a barrel', he says. Cruickshank then set up a new Callatoota in Denman, but a year later, in 2009, the then resources minister Ian Macdonald (now under investigation by the Independent Commission Against Corruption (ICAC)) approved a coal exploration licence which covered his new vineyard.[11]

This experience of being driven out has been replicated by a number of wineries, according to winemaker Ian Napier, head of the Hunter Valley Protection Alliance, which is now battling coal seam gas companies. 'I've had a gutful of promises made by mining companies which are not kept', he says.[12]

One of the most intense areas of mining is around the town of Muswellbrook. In this area, as in so many, some people claim the downsides of mining are justified by the jobs it provides. The local mayor, Martin Rush, questions whether small local towns like his own truly benefit in the long term from intense coal mining. Muswellbrook is undergoing a boom of sorts and has very low unemployment. But the plentiful jobs and high wages have not really increased the wealth of the town. This is because the high wages are accompanied by high prices. 'The purchasing power of high wages has been decreasing, because the cost of living has spiked', he says. A cup of coffee is expensive in a mining town because the cost of labour to make the coffee has to compete with the high wages of mine workers. The intensification of coal mining in places like Muswellbrook has 'overcooked' the local economy, he says.

One consequence of the over-cooked economy is that rents shot up 28 per cent in the last two years. Housing is scarce and in August 2012, 44 families were living in tents in the showground or at a local camping ground. This was the result of a 'disgraceful lack of planning' across the whole coal belt by the state government, Rush says.[13] He wonders whether Muswellbrook will prosper in the longer term. Many once great mining towns are now centres of poverty and disadvantage. In the Hunter this is true of Kurri Kurri and Cessnock, the home of the coal industry for 100 years. He points to other examples like Broken Hill, which once produced vast wealth for Australia and is now almost a ghost town. 'Where did the wealth go?' he asks.

How did the Hunter Valley become so degraded? According to John Williams, former NSW Natural Resources Commissioner, it happened through a 'death of a thousand cuts'. By this he means that each new mine on its own did not make a decisive change. But cumulatively they did. Planning based on environmental impact statements is a 'total failure' he says, and regional planning is needed. What's more, he warns, the 'Hunter is a vision of the future for other parts of NSW'. Wendy Bowman agrees. These days she meets other Australians whose communities are campaigning against the prospect of uncontrolled coal mining. She speaks to meetings in Northern NSW and Central Queensland. In different ways, she says, they all tell her the same thing: 'We don't want to end up like the Hunter Valley'.

To the north and west of the Hunter Valley are the Tablelands, where a series of coal megamines are operating, with more planned near the town of Mudgee. Local people near the existing mines have had their quiet rural lives turned upside down. Lance Batey and his wife Kate live near the village of Wollar. Escaping the city to the country with dreams of growing grapes, they are now trapped by the expansion of local coal mines.[14] After he tasted diesel in his tea, which he had made from water out of his rainwater tank, Lance decided to drink bottled water. Once every hour a 2-kilometre coal train thunders past, just 300 metres from their home. Once the nearby mines are complete the trains will run past twice as often. They are unable to sell their home, with real estate agents telling them, 'It's too close to coal mines'. Lance comments: 'There's no market for people like us. No one but a mining company would buy and they are not hurrying.' Many other neighbours left the area when faced with the expansion of the Ulan, Moolarben and Wilpinjong mines. Concerned residents of Mudgee recently held a 'Magical Misery Tour' through the area to illustrate the devastating impact of mines on local communities.

East of Mudgee is the productive grazing land of the Bylong Valley. The area was opened up for coal mining in 2008 and plans were laid for a series of four open cuts, plus an underground mine. When the mining company revealed these plans at a local meeting, residents were outraged. They had been told of only one pit. A campaign by local residents and inquiries by journalists about the details of the mining proposals ultimately led to a major investigation by NSW

ICAC, which is not yet finished. The inquiry is investigating whether the office of the NSW Minister for Resources, Ian Macdonald, passed on confidential information about lucrative coal-mining rights in the valley and whether he received or expected to receive money for granting exploration leases.[15] An ICAC public inquiry heard that Labor powerbroker, Eddie Obeid, along with some friends, had bought several farms in the Bylong Valley some time before a coal exploration licence was granted over the area. Shortly afterwards, coal mining company Cascade Coal agreed to buy three of these properties for four times their purchase price. As well, ICAC heard that the Obeid family took a substantial share of Cascade, which won the exploration licence and later agreed to sell their share for $60 million, though only $30 million was ever paid.[16]

Whatever the outcome of the inquiry, coal mining may well go ahead in the Bylong Valley, which will be devastating for local farmers such as Peter and Stuart Andrews. They have turned their farm, Tarwyn Park, once degraded and over-grazed, into a model farm. Peter says Australia has become 'mining mad'.[17] Stuart argues that, properly managed, the farm could be viable for 1000 years, instead of being sacrificed for short-term profit. It's not clear whether their farm will be mined or whether the mine will simply be next door. Either way, given the dust and degradation, they feel they may have to leave it. 'Who else is going to want to come here?' asks Stuart. The local Bylong Valley Protection Alliance has signs tacked on trees nearby that read: 'Food bowl, not coal hole'.

Mining the wide brown land

But if the farmers and residents of this area are upset, those further to the north are bound to be even more upset by a far bigger assault being planned by Big Coal. Over the Liverpool Ranges lies a vast flood plain feeding the Darling River to the west. Dotted with towns like

Gunnedah, Quirindi and Narrabri, the Liverpool Plains is one of Australia's richest farming areas. On these plains stands Kurrumbede station, once owned by the family of Dorothea Mackellar, whose poem 'My Country' talks of the 'wide brown land' and its 'warm dark soil'. Scientists refer to it as 'vertosol', a deep, naturally fertile soil infused with a particular clay mineral that retains moisture. Like the seams of coal that underlie the plains, the richness of the soil is an accident of geology. Millions of years ago volcanoes spewed forth a layer of rock called basalt nearby. When this clay-rich basalt breaks down and is washed onto the plains it produces this remarkable soil, which constitutes the most productive farming land in Australia. With its average rainfall of 650 mm the growing season is all year round. Wheat sown here yields up to 7 tonnes per hectare; elsewhere it's more like 3 tonnes. Farmland here sells for between $3000 and $6000 per hectare. 'If I had an unlimited supply of money to buy land to grow crops, this is where I'd invest,' said soil expert Dr Stephen Cattle from the Agriculture Faculty at the University of Sydney. 'This soil is the best we've got in Australia'. He says, with looming food shortages in a growing world, we should be wary of selling this land to overseas interests. This rich soil extends well up into Queensland and covers a number of areas under which coal is also found.

But such considerations don't count for much when there is a coal rush sweeping Australia. Less than a decade ago mining companies began to get very serious about mapping the vast coal beds underlying the Liverpool Plains. In 2006 BHP paid the NSW government $100 million for the right to explore for coal near Caroona, not far from Gunnedah, while the Chinese-backed Shenua Group paid $300 million to the state government. Previously such licences had cost around $10 million.

At Caroona, the mining companies struck a determined opposition. In 2008 some farmers began a blockade, refusing to allow BHP drill rigs on their property. They calculated that this was not mere exploration but a first step in a plan to gouge out giant open-cut mines. Farmers took a leaf from the book of environmentalists: they believe that their rich farming land should be treated like a world heritage area and be off-limits to mining.

Their cause was helped by a Senate Committee investigating the growth of the coal industry, which called for a total ban on coal-mining on the Liverpool Plains in December 2009. A successful legal challenge mounted by the farmers to BHP's right to enter property to drill exploratory holes brought an end to the blockade. But the victory was short-lived when the NSW Labor government undermined the court decision with a new law making it easier for mining companies to enter farms. At the time Caroona farmer, Tim Duddy, said:

> The government's reaction is not to uphold the law but instead to totally rewrite it to the benefit of the powerful mining lobby ... The message from minister Macdonald is clear: coal is king in New South Wales and let nothing stand in its way, including the law.[18]

Since then, the new Liberal–National government has put some restrictions on the extent of mining, aiming to protect the aquifers on the flood plain, but mining operations will still go ahead. Along the way, the Coalworks mining company bought the Kurrumbede Station and many other properties as a way of buying out local opposition.

The Saudi Arabia of the Southern Hemisphere

If NSW has the longest history of coal mining, then Queensland is the scene of today's most frenzied gold rush for the black combustible rock. On the surface of the land Queenslanders see bush-covered mountains, farms on the plains, a lacework of rivers and creeks and a string of coastal towns fringed by one of the wonders of the world, the Great Barrier Reef. However, Big Coal and its geologists see the Sunshine State in a different light. When they look at Queensland they see vast underground features they have called the Surat Basin, the Bowen Basin and the Galilee Basin. These basins, invisible on the surface, are giant geological features in which the gently tilted seams of coal, some up to 40 metres thick, along with deep reservoirs of methane gas, lie beneath farmers' fields, forests and towns in Queensland.

From these three giant basins (plus several smaller ones) big mining companies export 160 million tonnes of coal per year with more than one-third (60 million tonnes) going to Japan, mostly to feed its hungry steelworks, and another 20 million tonnes to Korea. In partnership with Big Coal, the Queensland government is in a race to increase coal production to more than 400 million tonnes per annum in 2025.[19] Over 30 new mines are planned on top of Queensland's 54 existing coal mines.[20] One of those mines at Wandoan (300 kilometres north-west of Brisbane) could be the biggest in the Southern Hemisphere and alone could produce close to 30 million tonnes per year with capacity to greatly expand. If Queensland finally reaches its desired goal of 400 million tonnes per annum (mtpa), it would have helped create nearly a billion tonnes of greenhouse gases annually. All of this would make Queensland the Saudi Arabia of Big Coal.

With obscene haste, its state government – once Labor and now Liberal–National – is undertaking a vast enterprise aimed at servicing Big Coal's plunder of the land. New railway lines snake out across the hills, along with power lines and dams, to supply coal's appetite for electricity and water. Highways are diverted and streams re-routed to suit the miners. On the Queensland coast, nine giant coal ports are planned or being built at a cost of over $7 billion. Overall, $19 billion will be spent to help speed up coal exports over the next 20 years.[21] Big mines and gas projects are routinely fast-tracked by being declared 'projects of state significance', thereby allowing the government to compulsorily buy land on behalf of private interests.[22] Proposed mines are rubber stamped and controls are lax. Two major projects illustrate this: in 2010 Cougar Energy began an experiment in burning coal while it was still underground to produce gas. The project, based in farm land near Kingaroy and worth $550 million, was closed down in early 2011. Tests found that the local groundwater was contaminated with benzene. The Queensland Ombudsman later found that the approval process and guidelines were sloppy. The company itself was allowed to set the 'safe' level of contaminants with no oversight from the Environment Department.[23]

The second case is different, but shows similar sloppy controls. In 2011 the Queensland Department of Employment, Economic Development and Innovation (DEEDI) was reported to have listed a non-existent $500 million coal-to-liquids plant as having completed an Environmental Impact Statement and gained approval.[23] The plant, which the department said would produce 75 million litres of diesel, also appeared on official federal government documents. But the plant does not exist, nor has it been proposed, let alone approved, according to New Hope Coal, its alleged owner.

Yet coal companies routinely defend themselves by arguing that their mines are subject to rigorous controls. The centrepiece of control is the production of an Environmental Impact Statement. But these routinely ignore the emissions from the combustion of coal, which account for the biggest environmental impact of all. In reality it's open slather for the coal-mining industry, whose partnership with the state government resembles a version of crony-capitalism based on the royalties which coal mining brings to the state government. All of this while Queensland's political leaders mouth empty phrases about being 'green' and climate-friendly. Crony coal capitalism has a rich history in Queensland. In the 1980s Sir Leslie Thiess made a series of loans to cabinet minister Russ Hinze which coincided with a Thiess company winning a coal project near Moranbah.[25]

In 2009 former cabinet minister Gordon Nuttall was found guilty of corruptly receiving $300 000 from the head of Macarthur Coal, Ken Talbot. Talbot's company was favoured by government decisions to grant mining leases and build railway lines.[26]

But not everybody is happy with the coal-friendly governments. When former Queensland premier Anna Bligh held a community cabinet meeting in Toowoomba in March 2011, she got a taste of the anger felt by many ordinary Queenslanders about her government's policies. At issue was the granting of exploration permits for coal which came within a few kilometres of the regional town of Toowoomba. Hundreds turned up for the meeting, including worried townspeople who discovered that the value of their homes had dropped overnight and local farmers who suddenly found themselves trapped in an uncertain future. A local newspaper described the scene at the Mount Lofty High School auditorium: 'Many at the meeting became emotional while addressing the Premier and their questions became confused and lost in their all-consuming frustration, providing an easy opportunity for the Premier to side-step their comments'.[27]

The weeks before had seen the eruption of local action groups in and around the town of Toowoomba. In the small town of Gowrie Junction, 700 people had attended a meeting called by local resident Lorraine Stern, who admitted she had never attended a protest meeting in her life before. 'We are just ordinary people trying to stop a mining invasion,' she said.[28] Her thoughts were echoed by another leader of the grassroots campaign. Jim Wiltshire formed the Toowoomba Coal Mine Action Group because of the threat of a mine near his home. He said, 'I was dumbfounded at the process [for creating an exploration lease]. It can affect so many people and can be so fundamentally flawed. We are not radicals, we just want to protect our home.'[29] Later, the protests spread to the city of Ipswich, where coal-seam gas drilling was proposed, as well as new mines.

The protests had some effect. In June 2011 the Queensland government vetoed an exploration permit near Toowoomba and in August it banned exploration within 2 kilometres of any town with more than 1000 residents. Even this largely token concession angered the mining industry, which complained that it would erode overseas confidence in Queensland. Not that it is likely to stop the re-opening of a coal mine at Ebenezer, south of Ipswich.

Beautiful town, shame about the coal

The origin of the fear that motivated this grassroots revolt can be found 50 kilometres up the Warrego Highway at two small settlements, Jondaryan and Acland. To many people these towns are a nightmarish vision of Queensland's future. At Jondaryan the mountains of stockpiled coal are taller than the Sydney Opera House. The wind blows coal dust onto homes, cars, gardens and into watertanks. 'This was just your typical country town over a decade ago', says Doris Lander, who has lived there for 25 years.

> Now there are constant trucks and trains coming through the town, which, along with the noise and dust from the mine, is making life unbearable. I've had nothing but respiratory problems with my family and my horses since the mines have been running.[30]

She is particularly worried about her 13-year-old son. Her neighbour, pensioner Glennis Hammond, moved to Jondaryan about three years ago.

> When I first moved here I remarked on what a beautiful town it is, but I said: 'Isn't it funny that you never see anyone in their yards?' They replied: 'Oh, you'll learn why soon enough.' Now I get sick in the stomach; I get severe lung problems.

The residents have written countless letters begging for something to be done, but nothing ever happens. The monitoring of the air quality is done by New Hope Coal, and the residents say the testing is only ever carried out on windless days.

The coal that is stockpiled at Jondaryan is dug from open-cut mines at the nearby town of Acland, or rather the former town of Acland. Today it has just one original resident left, Glen Beutel, who is holding out against New Hope Coal, the company that wants to mine the coal under his home. A former geologist and wildlife photographer, Beutel is a hero to many people in south-east Queensland. Indeed, his fame has spread worldwide. In June 2010 he was the subject of a *New York Times* article, which saw the fate of the 120-year-old town as 'a catalyst for pent-up anger over the coal industry's push into populated and farming areas'.[31]

Beutel grew up in Acland and the town holds many memories for him. His parents planted hundreds of trees in the town and helped create a memorial to those who died in the First and Second World Wars, Korea and Vietnam. But the trees have all been chopped down and the war memorial shifted to a nearby town by the coal company, along with graves from the cemetery. The town's other 100 or so residents all eventually sold their homes to the coal company. One former local dairy farmer, Mark Vietheer, explained why he eventually sold up. 'Would you like to farm next door to a mine? You would live in a dustbowl.' He estimates the company bought 100 properties. 'It caused great stress in our family. They wear you down in the end,' he says.[32]

Beutel does not resent the desertion of the town by its residents.

> With the coal mine getting closer and closer it is not surprising everyone has sold. Many felt intimidated into signing take-it-or-leave-it offers ... I'm the only homeowner left and I'm in no hurry to leave. Yes, I'm alienated, very much so. Sometimes I can't sleep at night worrying about the future. For decades the government of this state was run by the mining companies. It's no different now.[33]

In a concession to such feeling, the Liberal–National government temporarily stopped the Acland mine from being expanded, but is now looking at a revised plan for expansion.

The mine at Acland is in one small part of the Darling Downs, a rich farming area in southern Queensland. Like the Liverpool Plains of NSW, much of its fertile soil is deep, clay-rich and moisture retaining. One of the main objections from the Darling Downs farmers is that after mining the soil on which they depend will never be restored to its original fertility. At the Senate Inquiry, Greg Lane, representing the Queensland Resources Council, claimed that the mining and gas industries always left the country 'as we found it'. But can mining really restore the prime agricultural soil, which it regards as 'overburden', on top of the coal? Expert evidence at the Inquiry cast strong doubt on such claims. According to the Australian Society of Soil Scientists, miners have restored land for grazing and for bush, but nowhere in Australia have the clay-rich vertosols ever been rehabilitated to support grain and similar crops.[34]

Many people living around the small township of Wandoan have a similar scepticism for mining company promises about soil rehabilitation. Wandoan would be encircled by coal mines whose boundaries will be only 2 kilometres from the town. One nearby series of mines, Xstrata's Wandoan project, will be one of the biggest in the world, exporting 30 million tonnes per year, with the potential to expand to 100 million tonnes. Local farmer, Pat Devlin, says all that will be left after three decades of mining will be 'a big quarry'. He believes promises to rehabilitate the land 'will mean nothing'.[35] The mine owner, Xstrata, has bought the land of 40 local families, but several farmers have refused to sell, an action which has meant they ended up in the Land Court. They were joined by the environmental group, Friends of the Earth (FoE), which noted that the coal exported from Wandoan would contribute 0.15 per cent of total world emissions. 'This might sound like a small number,' said Dr Bradley Smith of FoE, 'but in fact it's the equivalent to the combined emissions of 72 countries in the world.'[36]

One farmer, John Erbacher, who refused to sell to Xstrata, said: 'I've got a mining lease granted right over my land. They plan an explosives store right in the middle of it and want a 600-metre buffer zone around.' Another farmer said the company deliberately emphasised such intrusions on landholders' properties in order to force them to sell. One official had warned that, 'It's going to be like the Hunter Valley of Queensland'. Anne Cameron, who, with her husband Brian, owns a property close to the proposed mine, is sickened by the onslaught. 'I can sympathise with the Australian Aborigines and how they were displaced with the arrival of Europeans.'[37]

Flying workforce, housed in dongas

Like many towns, Wandoan may have to get used to a fly-in, fly-out (FIFO) workforce who would work their 12-hour shifts based in labour camps on the outskirts of town. This aspect of the Queensland coal boom is quite unlike the booms of earlier eras. In previous years big mines in Queensland led to the growth of regional cities such as Mt Isa and Charters Towers. In coal-mining areas, until the 1990s state governments insisted that companies build housing, streets, schools, hospitals and recreation facilities if they wanted a mining licence. The result was towns like Moranbah, where an earlier generation of coal miners lived with their families, joining local clubs and sending their kids to school. All this has changed dramatically with the current coal rush. The coal giants now want to plunder the resources as quickly as possible and have pressured state governments to allow them to avoid the more costly, time-consuming construction of houses and towns near their coal mines. As well, the mining corporations are encouraged to do this by tax subsidies.[38]

Under the FIFO model, the workers are not given a real choice of whether they wish to live in a town near the mine with their families. Sky-high rents or the absence of houses make this impossible. Instead, many work their 12-hour shifts and live in mining camps made up of 'dongas', cheap metal huts that provide a bed and basic necessities. At best, the mining camps are soulless and boring, and at worst they are crowded with blokes, booze and brawling.[39] The isolation, long shifts, heat and loneliness of many of these camps takes its toll on workers. In early January 2011, a 55-year-old contractor who worked in a big gas project in WA lay dead in his 'donga' for more than a week before being missed.[40]

The FIFO method (with its partner, drive-in, drive-out – DIDO) is sapping the life out of towns such as Moranbah in the Bowen Basin region of Central Queensland. With its population of 10000 and a mining camp outside town with 4000 workers, the town of Moranbah is something of a case study. Long-time residents (many from mining families) say that for them FIFO means that the problems fly in and the benefits fly out. While governments celebrate the coal boom, paradoxically towns like Moranbah are finding life harder. The big wages of the FIFO miners are rarely spent in the town and the boom has seen rents on ordinary homes skyrocket to $2000 per week. This forces out anyone who is renting and not directly benefiting from a wage in the new mines, including retired people and employees of shops in town. Moranbah was also once a stop for tourists and inland travellers, who would stay at local motels. Today, rooms in the motels are permanently booked for weeks and months in advance by mining companies and so tourism suffers. In similar towns in the state, at one point mine operators had booked all available accommodation for up to six years, according to the National Tourism Alliance.[41]

Businesses in Moranbah can't find chefs, cleaners, checkout operators, dental nurses or receptionists because these jobs used to be filled by the partners of miners who once lived permanently in the town. Today the families of most miners are back in Mackay or

Brisbane. But it is the less tangible social effects on Moranbah's lifestyle and atmosphere that hurt the most. Sporting and service clubs, once the heart of the community, are finding it hard to function with the FIFO way of life, according to the Moranbah Traders' Association, a group of 65 shopkeepers who protested to a Parliamentary Inquiry in 2011.[42] Among the 'disastrous ramifications', they say, is that the town is losing its reputation as a safe place: changing from a stable community made up of families who put down roots, to one dominated by single men who come to work and feel no obligation to contribute to the community.

At the Moranbah Medical Centre local doctors and nurses are being swamped by coal miners with injuries. The numbers of medical staff, police and other service workers are allocated on the basis of permanent population, which radically understates the actual number of people in the community. In different ways these problems – high rents, labour shortages, pressure on local services and riskier streets – are increasingly being experienced in many regional coal towns like Moranbah. On the outskirts of a number of coal-mining towns, both in Queensland and NSW, the number of labour camps for FIFO workers is growing. Workers are often 'cooped up like chooks in a pen', said one MP and violent incidents are under-reported because they are policed by private security companies.[43]

It wasn't always this way. In the 1970s the conservative Bjelke-Petersen government insisted that the Utah mining company help build Moranbah as a condition of it being granted mining leases for Goonyella Riverside and Peak Downs mines. By contrast, the former Queensland Labor government worsened the situation. In 2011 it approved a 100 per cent FIFO workforce for BHP's new mine at Caval Ridge. The only condition was that BHP would have to build a meagre 160 houses at nearby Moranbah. The decision flew in the face of a spirited, year-long campaign by locals angry at the effects of FIFO on their town.

The new frontier

The coal rush in Queensland is far from over. As the world demand for coal continues to grow, coal companies are in a race to develop once remote regions such as the vast Galilee Basin in the western half of Queensland. While Australia as a whole currently exports over 300 million tonnes per year, the Galilee mines alone – according to their proponents' plans for nine mines – will export up to 330 million tonnes.[44] The area has attracted the outriders of Big Coal, like Clive Palmer and Gina Rinehart, as well as two key Indian businessmen. One is Indian multi-billionaire Gautam Adani whose Carmichael Mine (potentially the largest in Australia) could export 60 million tonnes a year to burn in Adani's power stations in India. In 2011 Adani bought part of the Abbot Point Coal Terminal for just under $2 billion and claims he will build a town for up to 12000 workers at his mine. The second is Alpha Coal, a joint project between another Indian billionaire, G. V. K. Reddy and Gina Rinehart (see Chapter 3). Their Alpha and Kevin's Corner mines hold almost 8 billion tonnes of coal and they plan to export 80 million tonnes a year to burn in GVK's Indian power stations. The third mega mine is Clive Palmer's 'China First' project, which plans to export 40 million tonnes per year with a capacity of up to 100 million tonnes. Largely supported by Chinese money, it would be one of the world's biggest, if it goes ahead. Other planned mines are owned by Chinese and Brazilian interests.

So huge and remote are the planned mines that they require hundreds of kilometres of new rail lines, as well as new coastal shipping terminals. Both Labor and Liberal–National governments have fallen over themselves to assist with this. In June 2012 the Queensland government approved two new rail lines to take Galilee coal to Dalrymple Bay, Dudgeon Point, as well as Abbot Point. The universal justification for government support is the creation of thousands of jobs by the mines. Yet two of the big Galilee projects – the GVK-Rinehart project and Clive Palmer's China First mine – have signalled that they might require overseas workers under Enterprise Migration Agreements.[45] More than that, it turns out that the mining boom is leading to a reduction in jobs in other sectors. Surprisingly, the evidence for this comes from the China First project, which commissioned an impact assessment from the AEC Group.[46] The assessment admitted that one of the adverse costs of the mine would be '3,000 jobs lost primarily across Queensland, particularly in manufacturing, agriculture and tourism'. This is because the high wages in the mines will tend to lift wages across Queensland, forcing cutbacks and in some cases driving some businesses to the wall.

Many of the 'jobs created' at these mines are also a mirage. Treasury officials told a Senate Inquiry that in a country like Australia when unemployment is low, 'it is not the case that individual industries are creating jobs, they are simply re-distributing them.'[47] (This also explains the need for foreign guest workers.) It gets worse. The economist Matt Grudnoff used the economic modelling provided in the China First assessment and applied it more widely. Taking the 39 advanced mineral projects in Queensland and using the assessment's modelling, Grudnoff found that for every two jobs created in mining, one other job would be destroyed.[48] The Queensland mining 'boom' would crowd out close to 20000 non-mining jobs, mainly in manufacturing, but also in tourism and agriculture. Finally, the 39 Queensland mines will keep the price of the Aussie dollar high, making manufactured imports cheaper and holidays for overseas visitors more expensive. In places like tourism-dependent Cairns, international visitors have fallen by 21 per cent.[49]

Many tourists come to Queensland to experience the Great Barrier Reef. But the reef and surrounding waters are under assault from coal-related industrialisation. Expansion of ports requires dredging on a massive scale, and this stirs up toxic mud on the harbour bottom. In 2011 large numbers of diseased fish were caught off Gladstone, and record numbers of dead dolphins, turtles and dugongs were found washed up on the shore. State government claims that dredging was unconnected to the fish kill were not accepted by independent scientists.[50] The dredging was the equivalent of dumping enough material to fill the Melbourne Cricket Ground 67 times on the inner reef area.[51]

Together the new and expanded ports will service coal ships, whose numbers will quadruple from 1649 per year to 6576, or more than 200 per week in 20 years (2032).[52] All this brings with it a high likelihood of collisions through tricky passages of the Reef. In April 2010 the Chinese coal carrier *MV Shen Neng 1* hit the reef off Rockhampton and dumped several tonnes of oil. In May 2012 a sugar carrier lost control and drifted over the top of Shark Reef, northeast of Cooktown. Next time we may not be so lucky and thousands of tonnes of coal may end up in pristine waters. But this is minor because the hundreds of millions of tonnes of coal passing through these waters each year will help damage the Barrier Reef in another way. The burnt coal will produce carbon dioxide in sufficient quantities to continue warming the seas, acidifying the oceans and ultimately ruining the reef. Because the Reef is a World Heritage listed site, its fate is of concern to the United Nations Educational, Scientific and Cultural Organization (UNESCO). Alarmed by threats to the reef, a UNESCO delegation visited Queensland in 2012 and reported that Australia had failed to properly protect it. The report stated: 'The scale and pace of development proposal appear beyond the capacity for independent, quality and transparent decision-making.'[53] Decisions had been reactive and driven by short-term economic benefits. 'Once approved, there is lack of enforcement of conditions attached to approvals ... and few penalties for non-compliance', the report said.[54]

An independent assessment of threats to the Barrier Reef is now under way, but no one believes that any key elements of the planned expansion in coal exports will be prevented by it. The initial reaction of current Queensland Premier, Campbell Newman, to the UNESCO report was the blunt statement: 'we are in the coal business'.[55]

So far, we've travelled from Australia's oldest coal mining region, the Hunter Valley, up through the Liverpool Plains, then north into the vast coal beds of Queensland. While this covers the booming black coal export industry it does not cover brown coal. This is found in the state of Victoria, which mines and burns Australia's' dirtiest coal for electricity. Its vast fields of brown coal (or lignite, the technical name) are reputed to be among the biggest in the world. The coal is easy to get to, often only 10–20 metres underground with seams up to 100 metres thick. The catch is that the energy value is much lower than black coal, in part because the moisture content varies from 50 to 70 per cent. This is 'young' coal and it is so dirty that a single power station, Hazelwood, is responsible for nearly 3 per cent of Australia's total greenhouse emissions. Until now, Victoria's brown coal mines have operated solely to provide fuel for the big power stations in the Latrobe Valley and nearby. But the temptation of coal royalties has meant that both Labor and Liberal state governments have pushed hard to develop an export industry. In 2009 the Brumby government was considering plans to allocate the right to mine brown coal to a company which planned to dry it and export 12 million tonnes of it per annum to India.[56] Other plans included support for producing liquid fuel and millions of tonnes of coal-based fertiliser – all justified by the assumption that 'clean coal' would soon be developed (see Chapter 6). The plans stalled under Labor, but were picked up with enthusiasm by the new Liberal government under Ted Baillieu, which replaced Labor in 2010. Soon after, a mining company began to explore for brown coal at Bacchus Marsh, just 50 kilometres west of Melbourne. It discovered an estimated 1–2 billion tonnes, which it wants to export to India.[57]

The Victorian Liberal government now plans to foster an export industry by issuing permits to dig up billions of tonnes of the dirty fuel. It was backed by former federal Resources Minister, Martin Ferguson, who says the Latrobe Valley could become a mining export hub on a scale similar to the Hunter Valley.[57] Liberal federal leader, Tony Abbott, has said, 'I want brown coal to have a future ... I think we should make the most of this asset, not close it down.'[59]

In many ways, Victoria's plans to export brown coal symbolise the hypocrisy of both major parties which claim to be concerned about climate change on one hand, but then on the other are happy to export carbon emissions around the world for a quick dollar.

Nor is this the end of the hunt for coal's quick dollar. The new frontier for coal mining in Australia extends to remote parts of Australia never previously considered as coal regions. For example, on Cape York a proposed $500 million underground mine would produce 1.5 million tonnes a year. The Wongai project, being developed with the agreement of Kalpowar Aboriginal Land Trust, could last for 30 years.[60] In WA a coal mine has been proposed for the Kimberley region, near the national heritage-listed Fitzroy River. One of the big attractions is the shorter distance from the WA port of Derby to the coal-fired power stations of India. The mine sponsor, Rey Resources, wants to transport the coal using road trains on the public highway. These plans have prompted a coalition of local residents, including Indigenous leaders and environmentally-minded people, to begin a campaign to stop the project.

All over Australia similar conflicts are growing between those whose lives are disrupted and others who are hungry for the coal dollar at any cost. The growing resistance to the coal juggernaut, which is steered by wealthy corporations and aided by compliant governments, offers our only hope that the strip-mining of Australia will produce its own counter-force.

3 THE BARONS' BOOM

'One of the benefits of global warming is there's not as many icebergs ...'
CLIVE PALMER, coal baron, climate sceptic, would-be Prime Minister and *Titanic II* proponent

If you take a walk down Queen Street, Brisbane, away from Fortitude Valley, and head around the corner to Edward Street, within the space of around 600 metres you will pass the flashy high-rise offices of a new breed of larger-than-life coal barons – Gina Rinehart's Hancock Coal, Clive Palmer's Waratah Coal, Peter Bond's Linc Energy and Nathan Tinkler's Whitehaven Coal. None were billionaires a decade ago and their rise to prominence provides some of the most visible evidence of the Australian coal boom of recent years. Their conspicuous spending and forays into politics, the media and professional sport have been impossible to miss. Their modest formal qualifications hint that boom-time cash is trickling down, that anyone playing their cards right can make a fortune. Their penchant for controversy has a polarising love-'em-or-hate-'em effect that draws the public into a 'Who wants to be a coal billionaire?' reality show of sorts. Though bizarre at times, we keep watching these larger-than-life contestants. As we will see, the number of Australians joining the 'nouveau coal-riche' list is rising fast, and yet they could hardly be less representative of the coal industry that is actually operating in Australia. In the main, it is not dominated by attention-grabbing billionaires, but by much older company names that are equally familiar to us.

Old coal

More than two-thirds of the black coal actually being mined in Australia comes from projects involving these five companies – BHP Billiton, Rio Tinto, Xstrata, Anglo American and Peabody.[1] Between them, they account for nearly 250 million tonnes of coal annually. Behind them are a dozen or so medium-sized players like Centennial Coal, Yancoal, Vale, Wesfarmers, Ensham, Whitehaven, New Hope, QCoal and Jellinbah Resources. Together, they account for about 70 million tonnes of mined coal. Then there's a raft of smaller companies, mostly spruiking large reserves but not yet producing.

Of the 'big five', the first four are multinationals diversified across a range of commodities; only Peabody is exclusively focused on coal. All these dominant companies have long histories in Australia. While BHP is now around 80 per cent foreign-owned, in the public's eye the 'Big Australian' tag remains. Perhaps it's because the company's fairytale rise since striking it rich with silver and lead in Broken Hill back in the 1880s is an important slice of our history. Perhaps it's the integral role played by the company and prominent Australians within it. As well as forging various Australian industries including steel, former managing director, Essington Lewis, was seconded as Director-General of Munitions during the Second World War. BHP Chairman, Harold Darling, meanwhile, wrote the first industry policy adopted by Sir Robert Menzies' Liberal Party.[2] For decades BHP has been one of the biggest companies listed on the Australian Stock Exchange (ASX), and has provided a procession of business leaders seeking to speak on behalf of industry as a whole. Rio Tinto is similar – by virtue of its various forebear companies it has deep roots in Australia, stretching back to and beyond the dominant period of Collins House (a collection of powerful miners who reigned supreme over Australian industry in the early 1900s).[3] Through its purchase of Mount Isa Mines and various other Australian assets, Xstrata (now GlencoreXstrata) can point to a longstanding Australian heritage. While gold rushes come and go, Mount Isa's rise in the 1920s was perhaps second only to Broken Hill (and perhaps Kalgoorlie) as the most enduring mineral success story in the country. Both Anglo and Peabody, meanwhile, have been prominent here since the late 1960s.

Thanks to their sustained presence and the depth of involvement these companies have had in economic and political development, many think of the coal industry as quintessentially Australian. Many would assume the Aussie barons that we are rapidly getting to know speak for the industry. In fact, none speak for the companies mining coal in large quantities – not Rinehart or Palmer, let alone Bond or Tinkler. These companies have different owners, most of whom we have never even set eyes on.

Sure, through their CEOs we might link a face to a company. Before their recent departures as the CEOs of BHP Billiton and Rio Tinto respectively, we saw a fair bit of Marius Kloppers and Tom Albanese in the media, warning us that Australia's competitive mining edge was at risk thanks to carbon and resource rent taxes.[4] Expressing these concerns in patriotic terms was unconvincing thanks to the detached foreign accents, but they acted like coal barons. However, with the exception of Ivan Glasenberg (new CEO at the merged GlencoreXstrata), those at the helm of Big Coal's dominant companies in Australia are dispensable hired help, as the recent turnover at BHP and Rio demonstrated. The CEOs are well known and well paid, and many have shareholdings beyond our wildest dreams. Company directors are also well remunerated, but often they are hired help too; there to open doors or inspire investor confidence, or to be the obedient nominee of institutions with large stakes in the company. But ultimately, it's others further up the chain who own the company; the people running it day-to-day are in relative terms, small fry. This is even more the case for the industry's lobbyists. They are well paid and well-connected, but their financial stake is infinitesimal compared with the industry's real owners. The power behind the industry lies with their owners, not the well-known men and women fronting them.

So, who are the faceless barons behind these companies? Which individuals, if any, have a controlling stake in the companies that are mining most of Australia's coal today? Answering this takes us down a Byzantine path, but it's worthwhile for what it reveals.

Banking on a dirty future

Perhaps not surprisingly, the multinational coal companies that dominate Australian production are overwhelmingly owned by institutional investors, most of them foreign-owned. Most obviously, banks have very large stakes. So, for example, the Hong Kong and Shanghai Banking Corporation (HSBC) owns 20 per cent of the Rio Tinto shares listed on the ASX. JPMorgan Chase holds 15 per cent; the National Australia Bank, 13.5 per cent; and Citicorp, nearly 5 per cent.[5] Much of the money comes from investors buying into a managed fund, who have little idea of exactly which stocks are chosen on their behalf.

In all, a few banks own over half of the Australian-listed shares in Rio Tinto, and this is not unusual. As well as owning shares in coal companies, banks assist their expansion in other ways. Various environmental campaigners have looked at which banks disproportionately fund coal mining, ports and power stations. Research commissioned by Greenpeace found that Australian banks provided around $2.3 billion in finance for new coal-mining projects in the last five years alone, and nearly $1.2 billion for coal port developments.[6] Although this did not include the large investment made by Australian banks in companies mining coal offshore, it reveals that, in spite of the many green accolades they spruik publicly, Australian banks are deeply enmeshed in the expanding production and use of the world's dirtiest energy source. Indeed, if the best-known Australian banks were people they would rank among our biggest coal barons.

Yet, the role of Australian banks pales in comparison with their overseas counterparts. A 2011 report by BankTrack, a European-based environmental finance group, uncovered a host of major banks worldwide that were each providing billions of dollars annually in project finance to new coal mines and coal-fired electricity. Since 2005, for example, JPMorgan Chase has provided over US$16 billion, while another seven banks have provided more than $10 billion over the same period (Citigroup, Bank of America, Morgan Stanley, Barclays, Deutsche Bank, Royal Bank of Scotland and BNP Paribas). Much of this investment is happening in Australia. JPMorgan Chase, Deutsche Bank, Barclays and the Royal Bank of Scotland (RBS) have provided over $7.1 billion to BHP Billiton, Anglo and Xstrata, all of whose coal investments are heavily concentrated in Australia. The analysis by BankTrack and others reveals much and makes a mockery of the climate-friendly marketing pitches of these banks.[7] It raises another question too – who owns the banks that are feeding the expansion of coal use?

The answers lie hidden beneath a money trail that is opaque by design. So, when we look at who owns the pillars of Australian banking we discover that the major shareholders are generally overseas-based investment vehicles. Some are banks and the same sorts of names mentioned earlier emerge again and again: among others,

JPMorgan Chase, HSBC, Citicorp, UBS. But who owns them? Well, often the trail leads to a foreign government. So, for example, the largest shareholder of UBS and Citicorp is the Government of Singapore Investment Corporation (GIC); the China Investment Corporation (CIC) has a large stake in Morgan Stanley.

It's the same when we look beyond banks at the fund managers dominating major shareholder lists of large coal companies (along with the banks). One good example is Pittsburgh-based BlackRock, which, as well as being the largest investment manager in the world, owns close to 5 per cent of Xstrata, Anglo, BHP Billiton and Peabody Energy (along with similar shares in dozens of smaller coal companies).[8] Look behind BlackRock and there you will discover various government-owned funds from Middle Eastern and Asian countries as significant holders. The Singapore GIC is a major shareholder, as is the Kuwait Investment Authority (KIA).[9] Perhaps the biggest surprise is Norges Bank, which manages the sovereign wealth fund established by Norway with oil revenue to ensure an ongoing national benefit as reserves decline. There is lots of high rhetoric from the Norwegian government about fighting climate change and how its sovereign wealth fund is being managed in an environmentally friendly manner. Yet, through Norges Bank's stake in BlackRock it has a sizable and largely unpublicised stake in expanding coal mining in Australia. At least with BlackRock you can find information. It's harder with other coal-laden investment vehicles whose convenient location in the Channel Islands and Caribbean tax havens seems intended to reduce transparency.

Faceless barons

Ultimately, we can pinpoint just a few living, breathing coal barons with billion dollar stakes in the companies that are mining most of Australia's coal. The most obvious is South African-born Ivan

Glasenberg, who owns just over 8 per cent of GlencoreXstrata and sits at number 2 on the *BRW* Rich List with an estimated fortune of $7.4 billion.[10] Following the merger between Xstrata and Glencore, with Glasenberg as new CEO, the reclusive baron has a more obvious stake in Xstrata's plans to expand coal mining in Australia (including possibly mining 100 million more tonnes of coal per year at the Wandoan coal mine in southern Queensland).

To a lesser extent, Nicky Oppenheimer, who has a 2 per cent holding in Anglo American and spent 37 years on its board, is another baron one might associate with a coal multinational. His family founded Anglo along with the De Beers diamond empire, and while today his stake in the company is much reduced, given that Anglo mines 40 million tonnes in Australia annually and plans to mine another 25 million tonnes or more, it's still a significant stake in our coal boom.[11]

However, beyond these and a couple of less notable exceptions the money trail takes us round in circles; rather than identifying powerful individuals controlling the show it reveals coal companies owned by banks owned by super funds owned by sovereign wealth funds owned by investment vehicles owned by other banks. It's like a long rendition of 'The old lady who swallowed a fly', without ever being able to find out who actually swallowed the fly.

One of the upshots of having an institution rather than an individual in control is that the main companies tend to be more machine-like – maximising returns for large institutional shareholders above all else. It's all about getting the best result for investors from around the world, and if that involves coal mining in Australia, so be it. Relying heavily on investors who are as fickle as their money is globally mobile fosters companies that are stateless, faceless and owe no real allegiance to Australia; all competing to maximise returns quickly to win and retain investment nods from large fund managers. It fosters decisions that are not about what's good for Australia, and even less about what's good for the environment. Since most investors with a stake in coal mining in Australia are unaware of it — with decisions made on their behalf by fund managers they don't even know — there's little investor pressure for companies to factor in the wellbeing of the planet into their decision making. The consequences for the global climate are a peripheral consideration as the pipeline of new coal projects planned by these companies in Australia reinforces. Collectively, Xstrata, Rio Tinto, BHP Billiton, Anglo and Peabody plan to add around 185 million tonnes of additional capacity by 2017.[12] That's equivalent to a 62 per cent increase in current coal exports in five years[13] — the emissions equivalent of trebling coal-fired power use in Australia.[14]

The coal nouveau-riche

Yet, as disturbing as this amoral march to environmental disaster is, the faceless barons we know could well be overtaken within a decade by a new breed of barons, with gargantuan plans to turbocharge the coal rush. Not so long ago, changes in the Rich List rankings happened gradually; there were the usual slings and arrows of economic fortune and personal foible, but it took time to build personal fortunes and the prominence and influence it brings. Pecking order changes were somewhat orderly and involved names with which we had become familiar over generations — Murdoch, Myer, Pratt, Packer. That's history with the rise of the nouveau coal-riche. In 2010, Australia's richest person (Frank Lowy) had just over $5 billion. Within 12 months Gina Rinehart was worth twice that.[15] Names like Lowy and Packer are being left behind thanks largely to coal. Just ten years ago, there were no coal barons in the billionaire section of the *BRW* Rich List. In 2011 there were four in the top ten, and dozens are making their way up the list. People who not so long ago worked as mining electricians, accountants and public servants, are now making barely fathomable coal fortunes. The rise of most of these people goes largely unnoticed. But, perhaps it's time we get to know these coal barons better. After all, if all goes to plan they will soon mine twice as much coal in Australia every year as the familiar multinationals, trebling Australia's contribution to climate change en route.

Broadly speaking, the 'coaligarchs' that are being spawned fit into two categories: Australians getting rich and foreigners enriching them to get richer still. First we will look at the template being followed by the Australians. Usually one person, or perhaps a few, have owned a large stake in, or in some cases all of, a relatively small company with a few coal leases. Because they are not household names like BHP or Rio Tinto, when they are bought out by large foreign-owned companies at many times the price they paid for their leases, their Eureka moment passes by unnoticed, barring mining industry analysts and stock watchers. With thermal coal fetching over $100 per tonne and coking coal nearly double that during the recent boom, coal-mining profit margins have made it one of the most lucrative industries in the country. Even a deposit that might ultimately produce 5 million tonnes a year has catapulted relatively small players into the front section of the Rich List. Once enriched, coal barons don't sail off into the sunset; they almost always stay on, trying to do it all again. Some barons are now on their second or third cycle of speculation, emboldened by seemingly unstoppable coal demand. Prices have now come down a bit, but demand marches upward. As a result, Australian coal barons are becoming 'world-class' barons, with Gina Rinehart (#36) and Ivan Glasenberg (#175) both on Forbes' list of the world's richest people.[16]

Those following the get-mega-rich-quick-from-coal game plan range from the larger-than-life conspicuous variety to those doing it very quietly. We're familiar with the first group. We hear their splashes in the media as they throw their weight around. We see them threatening to sue or bring down governments, donating money to political parties, and generally beating their chests. Yet few of us have heard of a legion of much quieter barons who are getting rich at the same breakneck pace, while studiously avoiding publicity. There's good reason to pay attention to the less visible barons, which might surprise many: they are generally the ones who are actually mining coal as opposed to talking about mining coal. As a rule of thumb, the more vocal and well known the coal baron – think Rinehart or Palmer – the less likely they are to be mining significant quantities of coal yet. More likely, gargantuan estimates of the net worth of these 'paper barons' are based largely on them on-selling a tenement, or the likelihood that they will eventually sell or start mining. To appreciate this better, let's look at the various paths some have followed to coal baronhood.

GINA RINEHART

Gina Rinehart is now one of the world's richest women, with an estimated net worth of US$17 billion.[17] Some of that wealth comes from the remnants of the West Australian iron-ore-based mining empire built, and then partly squandered, by her father, Lang Hancock. She secured that foundation after a bitter legal battle with her father's third wife, Rose Porteous, to whom Lang had directed a large share of his estate in his final days.[18] A large portion of Gina Rinehart's current wealth is based on estimates of the value of her 100 per cent ownership of Hancock Prospecting, which began exploring and acquiring coal tenements in the Galilee Basin of Central Queensland in the 1970s. Back then, Lang Hancock and Premier Joh Bjelke-Petersen championed the idea of building ports, and ultimately a transcontinental railway, to link Queensland coal with Pilbara iron ore to feed a new steel industry in the west. Together, the coal assets acquired by Hancock in the Galilee could produce 84 million tonnes of black coal annually; more than Australia currently burns in its electricity, iron and steel, and cement industries combined.[19]

For many years, the Galilee Basin was considered landlocked and highly speculative, but as demand for coal from Asia has grown, along with investor interest, the port and rail infrastructure required to make this coal accessible to the global markets has become a more realistic proposition. So, even though Rinehart's company doesn't yet produce any coal, apart from a small quantity produced from a test mine (promoted to the media as the Galilee Basin's 'first coal'), the value of its leases has spiralled.[20] In 2011 Rinehart cashed in some of this paper profit by selling over 85 per cent of her Galilee holdings to Indian company GVK for US$2.4 billion.[21] She retains a 21 per cent share in two of the three proposed mines (Alpha and Alpha West).[22] Her plans took a leap forward when GVK/Hancock won Queensland government support for a new terminal at Abbot Point, an allocation which received federal approval in late 2012. Whether she mines the coal or sells the remaining stake makes little difference; if the coal gets burned, one way or another Rinehart's net worth is likely to rise much further.

Like her father, Rinehart is a devout believer in the need to exploit what she sees as Australia's largely untapped mineral resources endowment – 'Go north young man', she implores us.[23] She is quick to promote her determination to exploit Australia's mineral wealth as patriotic, saying, 'While I have many investment options in a globalised economy, the place where I most want to create sustainable jobs is Australia.'[24] As proof of her commitment she says she spends her money on projects where others would buy private jets and yachts. She bemoans Australia's competitiveness against places like Africa where people will work for less than $2 a day.[25] In a recent magazine column, she suggested, 'If you're jealous of those with more money, don't just sit there and complain, do something to make more yourselves, less drinking and social time and more work time'.[26]

The Northern Australia focus is also pitched as lifting developing countries out of poverty. 'The world's poor need our resources

– do not leave them to their fate', she wrote in a widely circulated poem in 2011.[27] To advance this cause she set up and chairs a lobby group called Australians for Northern Development and Economic Vision (ANDEV). It has attracted the support of some of the biggest names in Australian mining. With near-separatist zeal, ANDEV promotes the idea of a vast exclusive economic zone in Australia's north, freed of regulatory red tape – quarantined from the usual taxes, environmental regulations and labour laws.[28] ANDEV also runs a 'North Australia Project' jointly in collaboration with Melbourne-based free market think tank the Institute of Public Affairs (IPA) (which some believe Rinehart funds, though the IPA won't disclose its donors).[29] Among other things, the IPA steadfastly opposes new imposts on mining projects ranging from the resource rent tax to the carbon tax, and argues that major resource projects should have to access cheap migrant labour. In magazine columns, YouTube videos and the aforementioned poem, Rinehart has advanced these sentiments, decrying policies she sees as responsible for Australia 'becoming too expensive and too uncompetitive to do export-oriented business'. In late 2012 she launched her book, *Northern Australia and Then Some*, which pushed the same line (prospective buyers are encouraged to buy it through the IPA).[30] It seems likely that Rinehart's desire to see Northern Australia's mineral wealth exploited is a genuinely held article of faith. However, her vested interest is equally genuine, and the self-serving reality is that the ideology, however real, helps to justify her plans to mine vast quantities of coal in the Galilee Basin.

A second self-serving article of faith is Rinehart's view that climate change is not driven by human activity. CO_2 is mere 'plant food' she says, keenly promoting the work of sceptics.[31] In 2011, for example, she 'recommended' that prominent denier 'Lord' Christopher Monckton deliver a lecture in her father's honour. Australian sceptic Ian Plimer (a strong supporter of her ANDEV initiative) has been appointed to the boards of two Rinehart-owned companies – one of them a coal company.³² Being so certain that climate change is natural and beneficial, that exploiting mineral resources quickly is good for Australia and essential to raising hundreds of millions out of poverty seems to embolden Rinehart to routinely conflate her commercial self-interest with the greater good. It perhaps also gives her the confidence and resolve to take various steps to translate her financial clout into media and political clout. Along with her ANDEV investment, Rinehart has bought a large stake in Fairfax Media, which many see as a strategy to steer the media empire's editorial agenda in a direction more in line with her agenda. In 2010 she funded a television advertising campaign by the Association of Mining and Exploration Companies (AMEC) opposing the mining super profits tax. In the same year she climbed on the back of a truck in Perth leading an 'Axe the tax' chant; an extraordinary step for someone who rarely ad-libs publicly and almost never grants media interviews. It all reinforces Rinehart's determination not to let climate change or anything else interfere with her plans to see the coal she owns mined.

CLIVE PALMER

Like Rinehart, Clive Palmer's vast net worth – estimated in 2012 at $3.85 billion, is largely based on yet-to-be mined coal reserves, the value of which has soared as the Galilee Basin looked more viable. By some estimates, Palmer's net worth has more than trebled since 2006. Some of the wealth comes from long-held iron ore investments in WA that entitle him to royalty revenue from Chinese company CITIC Pacific Mining, the timing of which is now subject to legal proceedings. However, much of it is based on the estimated value of his coal assets. Whether Palmer mines them himself or sells the reserves to someone else, his two proposed Galilee Basin megamines – Alpha North and China First – could produce over 80 million tonnes per annum.³³ The emissions generated when that coal is burned are equivalent to increasing by more than one-quarter the current annual emissions of global aviation (currently around 670 million tonnes).³⁴

Palmer has other coal leases in the Bowen Basin, and further north in the Laura Basin and elsewhere. If approved, the latter could open up large tracts of Far North Queensland to industrial-scale coal mining for the first time. Whether his grand plans to mine coal ever amount to more than talk is a matter of continuous debate. Generating some confidence is his partnership with Chinese government institutions that have deep pockets, Waratah has said the China Eximbank would provide up to 85 per cent of finance required;³⁵ the Metallurgical Corporation of China would oversee the development of the project, while China Power International would buy much of the coal produced at China First. The project would also mainly be financed by Chinese state-owned banks. Feeding scepticism about Palmer's grand plans, however, is the repeated failure to attract sufficient investor support to get his enterprise listed on the Hong Kong Stock Exchange.

Like Rinehart, Palmer effortlessly conflates the national interest with his private interest in coal mining, seeing it as a benevolent activity that raises people out of poverty. However, Palmer goes further than framing coal mining as patriotic and altruistic, declaring it also serves our national security interests. After all, he warns when referring to Chinese demand for Australian resources, 'It's better to talk than have them come here and fight us for our resources'.³⁶

Unlike Rinehart, Palmer constantly talks to reporters – a human headline whose gift of the gab and penchant for generating publicity have been enduring strengths. It helped the university drop-out make $40 million selling real estate on the Gold Coast in the 1970s and '80s, and to propel him into the thick of National Party politics at the perilous zenith of Sir Joh Bjelke-Petersen's premiership. Palmer became Queensland National Party spokesman during the 1986 state election campaign, which saw the Nationals rule in their own right (without the Liberals). He remained a senior player during the 'Joh for Canberra' push, when the enigmatic Sir Joh tested public support to make a switch to federal politics and a bid for the prime ministership (though Palmer maintains that he urged Bjelke-Petersen not to run).[37] Until Palmer's recent falling out with the merged Liberal–National Party – partly triggered by the Newman government's decision to increase coal royalties – he was a life member. As the intervention of Bjelke-Petersen's 92-year-old widow, Lady Flo, at the height of the dispute reinforces, it's hard not to think Sir Joh would be proud of Palmer's talent for toying with the media pack, or 'feeding the chooks', as he used to call it. One day Palmer is accusing Greenpeace and the Greens of being funded by the CIA, the next he's announcing his possible candidacy for federal parliament in the Treasurer's seat, or launching plans to build the *Titanic II* (which he says would be escorted by the Chinese Navy on its maiden voyage, and in which he would travel third class). Some of these claims are apparently serious, others later disowned as an obvious joke on the media pack. Sometimes it seems playful mischief designed to have a bit of fun with his growing audience, or to stymie and confound his many enemies. Occasionally enemies find Palmer less in the mood for jokes, and loose words suddenly land them in a protracted legal war against the man who reportedly once listed 'litigation' as a hobby in his *Who's Who* entry.[38]

Between the bluster, chook-feeding, litigation and the trappings of wealth, Palmer looks to encapsulate the lifestyle befitting a celebrity-star of 'Who wants to be a coal billionaire?'. He has his own office tower in Brisbane, and reportedly four homes, three private jets, two helicopters and a Bentley in the garage. There's a similarly impressive, if uniquely thorough, resumé to match. Well into his fifties, it features Palmer's schoolboy days sporting achievements: 'School Athletics Champion 1970, 1971, 1972', 'Club Rugby Highest points scorer 1971', 'Winner of numerous Surf Carnival Junior Championships 1971–1974'.[39] Even his abbreviated 150-word biography makes space to drop the names of famous world leaders recently met – from Vladimir Putin to the late Ted Kennedy.[40] In interviews, Palmer also drops the name of Mao Tse Tung (upon whose knee he claims to have bounced as a boy when he visited with his father, who he says was an 'unofficial representative of the Australian government').[41] Then there are the adjunct professorships from Bond and Deakin universities (to which he has contributed his time as a guest lecturer). Though he never completed a degree, Palmer uses the title of Professor, usually dropping Adjunct from the title for his profile on his company websites, and even his Twitter account.[42]

Yet, the big front and lavish trappings belie endearing generosity. Palmer has been a bighearted philanthropist to a wide range of causes – and an even more munificent boss (in 2010 he famously gave 55 Yabulu Nickel Refinery employees near Townsville a new Mercedes Benz each, while around 700 other employees received Fiji holidays for two). The Gold Coast's A-League soccer team, owned by Palmer until the team's recent demise, was allowed to use one of the company's private jets; another occasional user was the Queensland Liberal–National Party. So, while there's plenty of attention seeking, it's not all about Clive. It is, however, largely about mining coal, and environmental concerns don't alter that equation. In environmental terms, he argues, coal mining is pretty good, at least compared with coal seam gas: 'a mine is not something that's going to destroy hundreds of acres of land ... I don't think we have anything to worry about coal mining itself, because a mine only takes over a defined area where the pit will be'.[43] As for climate change, he doubts it is being caused by humans. As he told the audience of ABC Television's 'Q&A' program in 2011:

> I don't believe (human activity is) leading to global warming. That's what my personal view is, right. I do believe that there's been an increase in carbon emissions. Any small change to the atmosphere can make a major difference but, you know, as I said before natural carbon dioxide is about 97 per cent of the emissions.[44]

And in any case, as he said more recently at the New York launch of his *Titanic II*, 'One of the benefits of global warming is there's not as many icebergs ...'[45]

NATHAN TINKLER

Whether or not he remains a coal baron by the time this book is published, Nathan Tinkler's rise to baronhood still seems worth including. At 37 years of age, the former coal-mining electrician from Inverell in Northern NSW was (until recently at least) living the dream. Back in 2007 he borrowed heavily, and went close to losing everything, to put together a consortium that bought a Middlemount coal deposit for around $30 million.[46] Less than a year later, it sold to Macarthur Coal for $275 million in cash and shares. He emerged with 10 per cent of Macarthur, whose share price doubled before he sold out in 2008 for $441 million.[47] Then in 2010 through his investment company, Aston Resources, he bought the Maules Creek deposit from Rio Tinto for $480 million. Aston listed soon after on the ASX, valued at $1.2 billion. As the value of Tinkler's one-third share rose, he found himself catapulted to billionaire status. By late 2011, valued at over $2 billion, Aston agreed to merge with Whitehaven, a medium-sized miner focused mainly on the Gunnedah Basin, near Tinkler's Maules Creek deposit. Not content with this, he mounted a takeover bid for the combined company.

At the time of writing, Tinkler's takeover bid for Whitehaven looks doomed, the value of his holdings savaged by falling coal prices, his capacity to raise finance diminished. By some estimates Tinkler's net worth shrank by $2 million a day in 2012, with his Whitehaven stake (largely borrowed money) contracting from $1.2 billion to less than $600 million.[48] The creditors are lined up, the courts are involved, liquidators are checking solvency, and the prospect of bankruptcy is being mooted. The private jet has gone, as have hundreds of his race horses, and the NRL and A-League franchises he bought in Newcastle en route to it being dubbed 'Tinklertown' may also be in the balance. Although his wealth has shrunk as spectacularly as it grew, and his future is uncertain, his rise provides a good example of how coal barons are being minted.

Like Palmer and Rinehart, Tinkler is a larger than life character who attracts near constant media attention. His disdain for the media doesn't help; in 2010 he famously told one reporter from the *Sunday Age*: 'You're a f – ing deadbeat, people like me don't bother with f – ing you ... You climb out of your bed every morning for your pathetic hundred grand a year'.[49] His renowned temper has fanned media interest, as has his opulent lifestyle and his penchant for not paying creditors (including the Tax Office) until the very last minute.[50] Most recently, even as many creditors went unpaid amid speculation that he was in financial trouble, Tinkler found money for a new $15 million mansion in Hawaii, just months after making another mansion in Singapore his principal place of residence.

Though Tinkler's lifestyle resembles Palmer's, and his reluctance to deal directly with reporters resembles Rinehart's, he is quite different in important respects. First, there's no desire to play politics or proselytise his world view, there's no overt climate change scepticism to justify coal mining, and there's no pretence that it is all about rescuing people from poverty. With Tinkler, what you see is what you get – a risk-seeking, streetwise and lucky mining electrician turned billionaire who saw opportunities in coal and took them. The second thing that makes him different is that his company actually mines coal. Until recently he was just another 'paper baron', but the Aston merger with Whitehaven changed that. Not much, mind you. Given his 21 per cent stake in the 10 million tonnes of coal that Whitehaven expects to mine in 2012–13, his share amounts to perhaps a couple of million tonnes of actual coal being mined.[51] However, that number would rise fast if Tinkler were to succeed again against the odds with his Whitehaven takeover. If he fails, his demise will fuel the rise of other barons.

PETER BOND

Peter Bond is another to get rich by valuing coal not appreciated by others until much later. As a kid he says he used to 'lay [sic] under a tree, look up at the clouds and just figure out ways to turn nothing into something and that's the way I've been all of my life'.[52] The university drop-out and former truck driver from Camden on Sydney's southern outskirts saw an opportunity in the mid-1980s to make money hauling, hosing, raking and sorting coal that no one else wanted – 1000 tonnes of it taken from the Balmain coal loader and dumped at Kemps Creek in western Sydney. Four years later he had made his first million dollars. A former colleague from Bond's brief stint as a trainee metallurgist with BHP in Wollongong heard of his operations and got him a coal-washing contract at Appin in Sydney's far south. The pair bought a couple of mines, selling them later for many times what they paid. Bond's coal-washing business prospered and eventually extended to Indonesia, where in 2001 he endured a career-changing altercation during which a mine owner refused to pay and one of his men even drew a gun on Bond before eventually reaching a resolution.[53] Perhaps a little traumatised, Bond spent time lying under a tree in Fiji, reportedly seeking enlightenment. Then (and with the benefit of a 14-week 'platinum partnership' tour with self-help guru Anthony Robbins under his belt) he got back to the business of making something from nothing – only at a safe distance from the Indonesian forests.

In 2004 he saw an opportunity in more unwanted coal when he bought into Linc Energy, a struggling company focused on long understood but largely dismissed underground coal gasification (UCG) technology (combusting coal underground and harnessing energy in the 'product gas' that rises to the surface). Linc had assets offshore, but its prospects hinged on UCG taking off in Australia. It also held yet more neglected coal in the form of a Galilee Basin lease that might produce 60 million tonnes a year. As the region began to be taken seriously the value soared, and in 2010 Bond cashed in handsomely when Linc sold the lease to Indian company Adani. As well as a $500 million up-front payment to Linc (Bond holds around 40 per cent), Adani agreed to pay a $2 a tonne royalty. If the mine proceeds that royalty stream could be $3 billion.[54] Bond stands to benefit from another new thermal coal mine in the Bowen Basin – Linc Energy's Teresa project – potentially 6.4 million tonnes a year, most likely bound for power stations in China.[55] It's among 'non-core' coal assets that Linc wants to sell for the right price. Following a visit by Bond with Russian aluminium and oil billionaire Roman Abramovich to Linc's UCG operations in Chinchilla, speculation is rife that the latter is looking to invest in Linc. So, while Bond isn't yet in the coal billionaires' club (his net worth was estimated at $300 million in 2012), he's on track.

Bond's rise has followed the usual template – speculative investment on what looked like stranded coal, sold later to a foreign investor for many times the amount paid. Controversy doesn't shadow him as it does Rinehart, Palmer or Tinkler, but Bond is no shrinking violet. Linc Energy ads feature him clutching a chunk of coal against a bright green backdrop, with Bond speaking for the full 30 seconds.[56] In 2009 he starred in a Russell Crowe-narrated episode of 'Secret Millionaire', a reality television program that gives millionaires the opportunity to spend time undercover before revealing their wealth and helping battlers.[57] Next he starred in 'The Diesel Dash' – a celebratory 45-minute 'infomercial' by his 'independent' media company.[58] Masquerading as a documentary, it followed an epic cross-country trip from Chinchilla on Queensland's Darling Downs to Perth in a bright green Volkswagen Polo. It showcased Linc Energy's 'ultra clean diesel', with Bond in the driver's seat, once more doing most of the talking.

Like other prominent barons, Bond exudes confidence that mining coal is a worthy cause. Here there's none of Palmer's expressed drive to rescue millions from poverty or avoid a resources war with China. There's no outright climate change scepticism either. That said, at the height of the boom in 2010 Bond made it clear in relation to the proposed carbon tax that 'I don't think now's the time', saying that with '1 per cent or less of the world's CO_2 production' we should wait until 'India, China and the US "show their colours"'.[59] Even so, he seems convinced that his business is good for the environment, especially UCG. That this is a slightly less emission-intensive way of burning coal that would otherwise be left safely in the ground does not seem to bother him. This enables Bond to cast himself in the unpersuasive role of coal mining eco-crusader. On the brink of receiving billions in royalty income from perhaps the largest coal mine in Australia's history ('At full production, Linc stands to earn over $120 million per year', Bond recently told shareholders),[60] and with more Bowen Basin coal up for sale, probably the most environmentally friendly aspect of Bond's business is that he doesn't yet mine coal.

CHRIS WALLIN

It's hard to imagine someone becoming extremely wealthy in a more quiet fashion than Chris Wallin. He is the total opposite of all the coal barons that make the headlines, and yet with an estimated net worth of some $3.8 billion he's one of the most successful. His rise is all the more impressive given his background. After all, Gina Rinehart inherited a mining empire to work with; Clive Palmer had made more than enough money in real estate to allow him to retire in his late twenties (before getting into coal). Wallin, by contrast, had no comparable financial base upon which to build. Unlike Nathan Tinkler and others who have made millions from coal speculation, Wallin's path did not require extreme risk-taking. He took a different and more cautious track; an inside track. Before becoming a coal baron

Wallin was a humble bureaucrat, eventually rising to the position of Acting Chief Geologist in the Queensland Department of Mines. It was the culmination of a career spent mainly on identifying and mapping coal deposits so that mining companies could do what they do best. Presumably at some point the grass looked greener on the other side.

So, Wallin retired and set up his own company called QCoal and went about collecting coal deposits in the Bowen Basin. When one of them, called Coppabella, was sold on to Macarthur Coal, Wallin received cash from the sale, as well as ongoing royalties. By 2010, when Japanese steelmaker JFE Steel Corporation tipped in $600 million for a 20 per cent stake in another QCoal project called Byerwen (and agreed to take 2 million tonnes of coal a year from it), Wallin was flying.[61] The former public servant had morphed into a coking coal multi-billionaire, and this was before QCoal's production really even ramped up. If all goes to plan, the company will have a clutch of new coal mines up and running in the next few years, producing possibly 23 million tonnes.[62] Each year this will add roughly as much greenhouse pollution as hybrid vehicles saved in total worldwide between their introduction in the late 1990s and 2012.[63]

Wallin reportedly drinks better wine now than he used to, but there are no conspicuous displays of wealth; he still lives in the same house in The Gap, a typical Brisbane suburb, where he has been for 20 years. Nor is there any attention seeking: Wallin has no public profile, owns no sporting franchises or other showy trappings of coal baron success, and doesn't say a word about the environmental consequences of his business. He's the leading invisible baron, and plenty more follow in his quiet footsteps.

A conga line of would-be coal billionaires

Perhaps even more startling than the rise of the new coal billionaires is the much larger number of very rich men (and they are all men) following in their footsteps. Dozens have used the same coal-to-riches script en route to making hundreds of millions of dollars. Heading the list, at least in dollar terms, is Sam Chong, the Malaysian-born ex-mining engineer who came to Australia to study at the University of New South Wales in the early 1970s.[64] Few Australians would even have heard of Chong or his company, Jellinbah Group, which operates two mines in the Bowen Basin. The company's plans are relatively modest compared to the larger players, adding another 4 million tonnes of annual production by doubling output at its Lake Vermont mine. However, with a 23 per cent slice of the privately owned company, Chong's modest-looking share is already worth close to $1 billion, and it looks set to rise further. When he's not mining coal Chong buys office towers in Brisbane. The sky, it would seem, is the limit.

Next are Brian Flannery and Travers Duncan, each worth close to $600 million.[65] These long-time partners made close to $500 million each when Felix Resources, in which they held large stakes, was sold to China's Yancoal for $3.3 billion in 2009. Other Felix directors didn't do quite as well, but they walked away with hundreds of millions of dollars each. Today, Flannery and Duncan are directors and major shareholders with White Energy, whose main interest is in briquetting low-grade coal into an exportable product (yet another example of getting rich from coal previously considered unusable). Along with three other White Energy directors, the pair are also substantial investors (around 12 per cent each) in Cascade Coal, a company at the centre of anti-corruption hearings by ICAC in NSW.[66] Two other White Energy directors are on the Cascade board, as was Duncan until recently. White Energy had planned to take over Cascade, offering to pay $500 million in cash and shares for the company whose two significant coal exploration licences had cost it a mere $1 million. Had White Energy's shareholders approved the deal, it might have generated big gains for Cascade's shareholders (including an estimated $50 million each for Flannery and Duncan).[67] Instead, it went horribly wrong following allegations that one of Cascade's major investors was the family of former NSW Resources Minister, Eddie Obeid, and that his family also owned the land at Mount Penny on which Cascade's exploration licence was based, and that then Resources Minister, Ian Macdonald was offered $4 million through Obeid intermediaries as a kickback for setting up the licence on their property and running a public tender for it (against departmental advice).[68] All parties deny wrongdoing, the outcome of the ICAC Inquiry is not yet clear, and the Cascade deal currently looks highly uncertain. Whether it proceeds or not, however, White Energy might be sold on, propelling Flannery and Duncan into the coal billionaire's club.

Another example is Tony Haggarty, a former accountant who, along with a few other director/major shareholders of what was Excel Coal, made hundreds of millions of dollars when it was snapped up by Peabody in 2006 for $2 billion. Haggarty's slice was estimated at $220 million. Once again, there was no walking away when there were opportunities to double the dose; in 2007 Haggarty and his former partners jumped into Whitehaven Coal, the biggest miner in the Gunnedah Basin. Until recently, Haggarty served on its board as Managing Director. Like most of the barons actually producing coal, he eschews the public eye, telling the *Sydney Morning Herald* that he has no plans to buy a sporting team or private jet, and saying, 'I have no desire to be recognised in the street. It's a curse. You can't go back from it.'[69] Nor does he appear to have any desire to let climate change interfere with his plans, saying: '... it's the height of human arrogance to think that we understand climate and ... to think we can control it, I think that's utterly ridiculous ... Pollution is bad, I don't think anybody can argue with that. Whether CO_2 is the devil, I'm not sure.'[70] Notwithstanding a terrible year for the company following the merger with Tinkler's Aston Resources, Haggarty's 3.3 per cent share of Whitehaven is still worth an estimated $98 million. If Whitehaven's planned projects come to fruition, or the company is swallowed by a foreign 'big fish' like so many others, or if Nathan Tinkler pulls off a Houdini-like takeover himself, the value of Haggarty's investment is likely to soar and propel him into the big league.[71]

Someone racing him may be Jeremy Barlow, a mining engineer educated in Brisbane who spent most of his career advising his employers and clients about where best to invest in fossil fuels. Along the way he seems to have mastered the art of fattening up a business for sale. In 2007 his coal industry advisory firm, Barlow Jonker, was sold to another international consulting firm, Woods Mackenzie, for an undisclosed sum.[72] In 2010, the coal seam gas company he had co-founded (originally CH4, later Arrow Energy) was sold to Shell and PetroChina, netting Barlow an estimated $40 million plus over four million shares in a spin-off company called Dart Energy. Along the way, the career analyst, who had spent much of his time watching coal markets, quietly shifted focus from gas to coal via a private company called Bandanna Coal, which was accumulating leases in the Bowen and Galilee basins. In 2009 Barlow and his co-owners took that company onto the ASX by merging it with an existing listed company (Enterprise Energy) that was having little success in oil and gas. Over 300 million shares were issued, of which the Bandanna Coal shareholders got 95 per cent, and Barlow and his colleagues took over the board of the new Bandanna Energy. Restrictions prevented them from selling out for two years, but Bandanna made it clear pretty quickly that their objective was to sell to the right bidder.[73] The company doesn't yet produce, but it claims the largest thermal coal inventory in Australia and plans to extract 15 million tonnes per annum from yet another huge mine in the Galilee Basin (the South Galilee Project).[74] Though he recently retired as chairman of the company, through various entities, Barlow still has nearly 120 million shares.[75] Should the share price return to its 2011 high of $2, and the right buyer emerge, his stake could be worth $240 million. Since most of his shares appear to have been acquired when the price was less than 10 cents (plus a few in 2011 when it fell sharply from $2), it could be a spectacular capital gain. The consequences for the atmosphere are the main downside – Bandanna's South Galilee Coal Project alone would generate CO_2 equivalent to over 8 million cars.[76] The nearest hint of any qualms over this is Barlow's quiet interest in biocarbon capture and storage. In 2009, as he was taking Bandanna public, he was also participating in a soil carbon tour of Victoria; not out of concern about climate change, it seems, but reportedly over his concern for 'land degradation and the possible impact on the prosperity of future generations'.[77]

Like Barlow, whose coal baronhood came via coal seam gas, quite a few barons have dived into coal from other sectors; it was just too good not to join in. One example is Andrew 'Twiggy' Forrest, who we think of as an iron ore baron, but whose subsidiary, FMG Resources (Fortescue Metals Group) has quietly bought a range of coal exploration permits in Queensland near Emerald, Nebo and Dysart. As well as joining the local rush, FMG Resources recently sought to join the Mongolian coal rush, and took out coal leases on the west coast of New Zealand near Greymouth.

Norman Seckold is another. He made a fortune through Boligni Gold, bailing right before it collapsed. Following that, he took a stake in Cockatoo Coal (chairing its board until recently), a company sitting on 2 billion tonnes of coal, with plans to produce 10 million tonnes per annum within five years (the emissions equivalent of five large new coal-fired power stations).

Some of Travers Duncan's partners in White Energy and Cascade Coal have also dropped in from other careers – John Kinghorn is best known for his roles with Krispy Kreme Donuts, Allco and RAMS Home Loans; John McGuigan and John Atkinson were partners at legal firm Baker & McKenzie (which advised Cascade shareholders on the deal with White Energy).[78] Tony Haggarty's replacement as Whitehaven CEO, Paul Flynn, used to work as a partner at Ernst & Young.

However, by far the best example of a baron who couldn't resist the coal rush is Kerry Stokes. Most Australians think of Seven Group Holdings (SGH), 67 per cent owned by Stokes, as a media company, not a coal company. However, the seemingly unlikely marriage of these endeavours is well and truly under way. SGH owns the WesTrac business which holds the franchise for Caterpillar, the world's largest provider of mining equipment. As the WesTrac website states, 'In New South Wales, WesTrac Group's operations are levered to the state's large and growing coal mining sector'.[79] This leverage has delivered record results.[80] In 2012 the company cited expansion in the coal industry, particularly in the Hunter Valley, for an 82 per cent increase in product sales by WesTrac in Australia – everything from three-storey high trucks to draglines weighing up to 7500 tonnes with booms over 130 metres long, wielding shovels carrying 100 cubic metres at a time.[81] Not content cashing in on the coal mining here, Stokes's company also owns WesTrac China, which provides heavy equipment to the mining sector in that country's north-east, with a strong emphasis on the coal rush in Shanxi Province and Inner Mongolia. SGH also has close to $250 million invested in the Agricultural Bank of China, the world's fifth largest financier of coal-fired power companies (€1.5 billion since 2005), and fifteenth biggest financier of coal mining (over €1 billion since 2005).[82] So whether the mining happens here or in China, Kerry Stokes wins big. Beyond this, he also has a large stake in HRL Limited, which runs a small power station in Victoria and had grand, but unfulfilled plans to build a much larger one.

That Stokes is a billionaire is well understood by most Australians, but few would link his fortune with increased coal use. Therefore, when not so long ago he warned that Australia needed to take advantage of Chinese demand for Australian resources or risk China looking elsewhere, it perhaps looked like a disinterested captain of industry exercising economic patriotism.[83] Likewise, when he recently said, 'What is important for Australia and for the future is that we continue to ship the resources', dismissed the carbon tax as a wealth distribution mechanism, and referred to coal-fired power, 'which we'll be using in this country for at least 25 years'.[84] In fact, the mining-dominated WesTrac business now generates three times the earnings of Seven's extensive media investments.[85] So, although it may not get spelt out this way in Channel Seven's news coverage any time soon, the company's future and that of Stokes's current fortune of $2.79 billion depends to a large degree on the continued expansion of coal mining, both in Australia and China.

Offshore barons

As Kerry Stokes's stake in Chinese coal mining suggests, not all Australian coal barons are striking it rich in the Galilee Basin, or even in Australia. Much of the enrichment happens elsewhere. With the strong backing of Australian government agencies like Austrade, the coal industry equivalent of cashed-up missionaries now trot the globe offering coal-fired financial salvation/Nirvana to developing countries susceptible to conversion.

Michael O'Keeffe is one example. In 2004, with partner Steve Mallyon, he set up Riversdale Mining, which focused on two large coking coal prospects in Mozambique. O'Keeffe's view was that coal demand for power generation had a bright future, so they chose Africa, partly, he says, because there was less competition. Notorious for its lack of infrastructure and the logistical challenges of barging large quantities of coal down the Zambezi River for export, the former MIM and Glencore executive was taking a big punt when he bet on Mozambique. You get the sense he knew it, and that he likes big punts. O'Keeffe has built close ties with Bart Cummings over the years as a keen horse-owner. He and his horse syndicate organising partner, Veronica King, bought a gelding called Moatize, named after the region in Mozambique where O'Keeffe hoped to win big on coal.[86] When Californian hedge fund manager Passport Capital bought into Riversdale in 2008, O'Keeffe and Passport's chief investment officer, John Burbank, sealed the deal, buying a horse together. In 2008, the horse named Moatize ran the Melbourne Cup, and the former North Queensland boy who now owned a pub in the middle of Townsville's night club strip promised 'we'll be shouting the bar' if the horse wins.[87] It didn't, but O'Keeffe got to keep his money — a sign of things to come.

Meanwhile, over in Moatize, Mozambique, his much larger gamble was really paying off: O'Keeffe's project attracted investment from a range of steelmakers, including India's Tata and China's Wuhan Iron and Steel, each of which took large stakes in the company. The company's share price rose from 20 cents to over $8.[88] It went into rapid reverse with the 2008 financial crisis, of which O'Keeffe says, 'we were all sweating', but Burbank stuck with Riversdale, and soon afterwards the upward march headed to new heights.[89] Then O'Keeffe's ship came in when Riversdale was swallowed by Rio Tinto in 2011 for US$4 billion, a purchase the latter would regret when it resulted in a $3 billion write-down and the departure of its CEO, Tom Albanese. According to news reports, O'Keeffe had considered staying on to work with Rio Tinto, but was put off by the company's austere Brisbane office. Instead, with an estimated fortune of well over $50 million, courtesy of the value of his nearly four million shares (excluding options) at the time Rio made its fateful purchase, he took the money and ... you guessed it ... sought to do it all again. O'Keeffe and Mallyon are now working on what they call Riversdale II – focused on coking coal tenements in North America. The two most prominent options being explored are a large deposit in Alaska (owned by the Alaskan Mental Health Trust), and another in Alberta, Canada, whose convenient location close to a rail line connected to coal ports in British Columbia promises to be cost-competitive with Australian coal mines. Mallyon says, 'We have always thought we can do better than Riversdale Mark I. Mark II will hopefully be bigger and better'.[90] Perhaps with an eye on the egg left on Rio Tinto's face after logistical hurdles in Mozambique contributed to the big write-down, he stressed to *The Australian*, 'We are focused on replicating Riversdale Mark I, but slightly smarter this time in that the projects are in first-world countries and adjacent to infrastructure'.[91]

Barons behind the barons

Whether offshore or back in Australia, becoming a small fish worth being swallowed by a big fish for spectacular profit is a process being replicated over and over again. Peruse the boards of the smaller coal mining companies – those that have been on-sold (e.g. Centennial, Excel, Felix), and those still around (Whitehaven Coal, Gloucester Coal, Aquila Resources, Cockatoo Coal, Linc Energy, Ambre Energy, Tigers Realm, Coal of Africa, Coalspur and a host of others) – and the same story is being written, starring many of the same characters. However, there's a flipside to all this: the foreigners paying vast sums to create Australian coal barons. This includes multinational mining companies we are familiar with, such as BHP Billiton and Rio Tinto, and other companies mainly owned by large foreign coal barons (and in some cases, foreign governments). Yet, what's most striking about the pipeline of coal projects proposed in Australia in the next decade is that it is not dominated by the household name multinationals, but by other companies with no real history in Australia, and in which individuals do hold a very large share. Let's look at a few of these barons behind the barons.

GAUTAM ADANI

Most Australians have never heard the name Gautam Adani, but his name will soon be familiar if he succeeds in building what could become Australia's largest coal project – a mega-mine producing 60 million tonnes of coal a year in the Galilee Basin. With a net worth estimated at around US$4 billion, Adani is India's sixteenth richest man. The 50-year-old self-made billionaire dropped out of his Bachelor of Commerce degree at Gujarat University and became a diamond sorter before going into a plastics business with his brother. The timing was good as Indian trade restraints were about to come down, enabling him to build a thriving import–export business. Adani expanded into infrastructure and power generation, but pivotal to his success has been the establishment of a huge port and Special Economic Zone (SEZ) at Mundra in India's north-west. It has become a big money spinner, with his existing coal-fired power station at Mundra (2 gigawatts, with plans to more than double that) and his port terminals and private railway feeding various facilities operating in the SEZ.

The opportunity for Adani to expand his power production business is enormous in a country where electricity demand is spiralling. India approved 75 new coal-fired power stations in 2010 alone, but domestic coal production simply isn't keeping pace with demand. New coal mines are hampered by a combination of creaking infrastructure, bureaucratic delay, environmental concerns and community opposition. So, although India has some of the world's largest coal reserves, it has a coal shortage. In 2012, thanks in part to an unmet spike in electricity demand, India suffered its largest electrical blackout in history, affecting 600 million people. Adani has felt the domestic coal shortage. His plans to build a 3.3 gigawatt power station in Maharashtra were supposed to involve a local mine, but concerns about tiger conservation got in the way. He explained, 'One morning, suddenly, the ministry of environment and forests barred our coal mine ... I thought okay you have a concern with the tiger sanctuary, let me work out other solutions'.[92] Tigers weren't the only reason Adani has opted to focus on coal supplies from abroad. With lead times for new coal projects in India blowing out to a decade or more, and controls on coal pricing another impediment, offshore supplies look more reliable. So Adani, and other large Indian coal users, have determined it is quicker, easier and safer to mine coal in Indonesia and even Australia, than it is to rely on Indian coal and infrastructure. As a senior Adani executive noted in a 2011 presentation in Brisbane, approval processes for a new mine in India can take 15 years before land acquisition is complete and mining can start; more than twice as long as that anticipated in Australia.[93] Adani contends that it's not about making money: 'Money doesn't drive me. I like challenges where you feel you are part of nation-building. I could have created many different businesses but I feel more satisfied when I create something that can be a part of the India journey'.[94]

This is the context of Adani's decision to spend $10 billion building the Carmichael coal mine in Central Queensland's Galilee Basin. The scale of the project is gargantuan, nearly matching the entire existing Australian coal production of BHP Billiton. Back in India it involves the world's largest coal import terminal at Adani's Mundra port, and it's central to his plan to have 20 gigawatts of coal-fired capacity in place by 2020 (equivalent to about two-thirds of the current total production in Australia).[95] Once used, the coal from the Carmichael project alone will generate roughly 160 million tonnes of CO_2 annually – roughly as much as is currently emitted within Queensland's borders every year, enough to erase all the projected emissions savings from Australia's renewable energy target, its carbon pricing scheme, and the $10 billion Clean Energy Finance Corporation.[96] Yet, crazy as it sounds, Adani will be rewarded with United Nations (UN)-sanctioned carbon credits for burning the coal he plans to mine in the Galilee Basin. That's because some of his new coal-fired power stations (including units at Mundra, which at nearly 4.7 gigawatts will be among the world's five largest) are more efficient than most of what exists in India (which isn't hard), and so qualify to earn carbon credits under the UN's Clean Development Mechanism.

If all goes to plan, Adani will export from either Abbot Point north of Bowen (where he has already purchased a terminal) and/or Hay Point, south of Mackay. The project would involve hundreds of kilometres of new railway. As he has done back in India, Adani's stated preference is to own and operate as much of the process from the coal mine through to the power stations – his own ports, railways and trains. He has made a prosperous habit of building everything himself and vertically integrating businesses. Relying on others perhaps comes less easily to a man who was kidnapped and held for ransom in 1999 by Mumbai gangsters for unknown reasons, before ultimately being released for 30 million rupees. Reports of private security forces intimidating unions, farmers and environmentalists have become hallmarks of 'Adani-towns' in India.[97] In Australia Adani has been on a charm offensive and enjoys the enthusiastic backing of state and federal governments. In December 2012, for example, he hosted Queensland Premier Campbell Newman and federal Resources Minister Martin Ferguson in India. Having been ferried around in Adani's private jet to see one of the billionaire's power stations, and to take in the Adani-sponsored Ozfest, the three men exchanged compliments for the media and posed happily for the cameras.[98] So far, the plan is on track.

G.V. KRISHNA REDDY

Not far from where Adani plans to mine, another wealthy Indian businessman is looking to cash in on Galilee Basin coal. Worth an estimated $1 billion, G.V. Krishna Reddy (otherwise known as GVK) is one of India's 50 richest people. The 68-year-old got a taste for the infrastructure business working in his uncle's construction company, building dams and canals. After a stint abroad he returned and took up particleboard manufacturing. In 1992 his big step was building the first privately owned power station in India. According to his son, Sanjay (now a senior player in the business), this was the turning point that encouraged GVK to think big. Since then, the company has diversified into hotels, road-building, biotech and much else. The tennis-mad Reddy has even diversified into sports in a sense, with GVK sponsoring promising players reportedly as young as eight years old, as well as Sania Mirza, India's top-ranked player. However, Reddy's main game is resources and infrastructure, and in 2006 many spectators were surprised when GVK prevailed against more highly fancied competition to win the contract to build the expanded Mumbai airport.

Excelling in the coal-fired electricity arena is also a big part of his plan. While recent coal shortages in India have delayed the timeline, GVK plans to build over 5 gigawatts of additional coal-based thermal power near Hyderabad in India's east. According to the company, half of the coal would come from Indian mines and half would come from its share in the new mines in the Galilee Basin bought from Gina Rinehart. Like Adani, bureaucratic delay and unreliable Indian supply encouraged GVK to look offshore. Already, gas shortages have left some of Reddy's gas-fired power stations running at only 25 per cent capacity, and as GVK Power recently noted of the state-owned Coal India,

> Even if you sign fuel supply agreement with Coal India, you will get only 60 per cent of the allocated coal. It makes no sense for anybody

to go ahead with new project [sic] ... It is better for us to wait for Hancock coal mines to become operational and get the coal to India and build projects ... the moment coal is available, we will ship it from Queensland.[99]

Unlike Adani, GVK seems less concerned about building his own infrastructure. Instead, he has agreed to jointly develop the port and rail infrastructure with Aurizon, which, with the Queensland government as its largest shareholder, minimises political risks by giving the state a bigger slice of the action. Importing their own labour is also on GVK's radar; Sanjay Reddy calls the option of importing migrant guest workers into Australia 'good insurance'. As with Adani, the implications for the climate are huge, based on what GVK plans to do. GVK mines in the Galilee Basin would produce 84 million tonnes of coal per annum, more than Xstrata currently produces in Australia, adding an extra 225 million tonnes of CO_2 annually once the coal is burned. To put this into perspective, as part of its 'Clean Energy Future' plan, the Gillard government said it wanted to close down 2 gigawatts of Australia's dirtiest coal-fired power, a measure that might save roughly 20 million tonnes of greenhouse pollution annually. GVK's mines, which have the enthusiastic backing of the Gillard government, add 11 times as much CO_2 as that pledge (since abandoned) might have saved.[100]

Like Adani, Reddy's coal-based expansion is also being rewarded with eligibility for UN-sanctioned carbon credits on the basis that some of his new power stations feature 'supercritical' coal-fired generating technology, which is more efficient than the more typical 'subcritical' variety. It raises the farcical possibility that Australian electricity generators might soon 'offset' emissions from burning coal with carbon credits purchased at new Indian coal-fired power plants burning even more of the same Australian coal. Far from expressing any contrition about all this, GVK paints itself as a green company. Laced in lots of pictures of green leaves and gently cupped seedlings, the company's website explains how it is 'reducing carbon footprints', that 'Sustainability remains a constant in all of GVK's projects and initiatives' and that 'one of the highest priorities during project implementation and planning is safeguarding the nature [sic]'. At GVK, sustainability is 'built-in in the processes'. A rapidly growing carbon footprint is also built in, but this seems to cause the company no discomfort.

Reddy's political reception in Australia is unlikely to cause him discomfort either. Both sides of politics have keenly encouraged Reddy to join the coal boom down-under. After an extended courtship of Gina Rinehart prior to buying a controlling share in Hancock's coal mining projects, she flew, along with federal Deputy Leader of the Liberal Party, Julie Bishop and Nationals Senate Leader Barnaby Joyce, to Reddy's daughter's extravagant three-day wedding festivities involving 7000 people. Labor Resources Minister Martin Ferguson was invited but declined to attend, stating, 'While valuing my relationship with them I did not think it appropriate to attend the wedding'.[101]

Probably more important from Reddy's standpoint would have been Ferguson's strong support for the projects in which GVK is invested. In 2010, Ferguson unveiled a plaque at a test mine established at the Alpha coal project, calling its proponents 'trailblazers', celebrating the possibility of another coal-producing region to rival the Bowen Basin and the Hunter Valley.[102] Then, in 2012, Ferguson's colleague, Environment Minister Tony Burke, approved the GVK mine under 'strict conditions' in which, farcically, the emissions from the coal to be produced don't even figure.[103] Reddy said he was delighted with the federal government's decision.[104] Well he might be.

YAO JUNLIANG

If Yao Junliang has his way the Galilee Basin won't be left to the Indians. Being the richest man in Shanxi might not mean much to many Australians, but as the province producing most of China's coal barons, it is becoming increasingly relevant. Yao and his family are one of China's 100 richest families and the 61-year-old is Chairman of Meijin Energy, China's largest commercial producer of coke. According to a profile in HSBC's *Week in China* publication on the country's tycoons, Yao spent his early years as one of Chairman Mao's 'barefoot doctors' working at rural health clinics, before he and his brother borrowed enough money to start renting out a couple of trucks, through which they got into hauling coking coal to nearby smelters.[105] It's a far cry from the conglomerate controlled by Chairman Yao today, and from his personal fortune, which has reportedly increased by roughly a third in the last two years to US$1.3 billion. Wholly owned by the Yao family, Meijin Energy is low profile in Australia so far, but that's likely to change. The company has a privately owned subsidiary called Macmines Austasia, and through it a mining lease in Central Queensland at Yarrowmere North, 300 kilometres west of Mackay. It plans to build a 60-million-tonne-a-year coal mine, which it is calling the China Stone Project. Most of the coal would go back to China to feed new coal-fired power stations. So far, state-owned enterprises back in China have agreed to buy 30 million tonnes per annum. Government assessment processes are still under way. If it proceeds, and depending on what happens with other Galilee Basin projects, Meijin's mega-mine could be the largest in Australia.

Once burned, the coal from China Stone would result in over 162 million tonnes of CO_2 annually – roughly as much greenhouse pollution as all the power stations burning black coal in Australia combined.[106] Not all of those emissions will occur overseas either. Anticipated electricity demand at the mine itself (primarily to power draglines and other equipment) is so great that Meijin says it wants to build its own brand new 800 megawatt coal-fired power station on site.[107] On its own, this new power station would comfortably erase roughly half the emissions being saved by all wind- and solar-powered electricity generation in Australia.[108] There's no word yet on whether the UN will reward Meijin's customers back in China with eligibility for carbon credits, but there is every reason to expect that might happen.

In spite of China Stone's likely contribution to climate change, neither the Queensland or federal governments have so far raised any objection whatsoever to the Meijin plan. The company's good connections haven't hurt it. ASIC records show that Macmines has just three directors, the first two being Yao himself and 30-year-old Ms Yao Jinli (also referred to as the company's CEO).[109] The third director is Geoff Dickie, appointed less than six weeks after he finished up, not just as Queensland's Deputy Coordinator-General, but as a Deputy Coordinator-General in charge of Project Assessment and Attraction (an agenda dominated by coal-related activity).[110] There's no suggestion here of wrongdoing, but it certainly looks an intriguing dynamic – for the Office of the Queensland Director General, knocking back the China Stone project's approval application means knocking back a former colleague. For Chairman Yao, it's a comfortable position.

Whose rush?

So, what are we to make of the coal barons and their seemingly inexorable rise? The expansion plans of the familiar companies that built Australia's coal industry, and the new generation of coal-fired billionaires, reinforce what many want to believe about the mining boom: that it is being built by Australians, largely owned by us, run by us, and that its economic contribution is shared among us. We imagine a hungry China and India that cannot do without our coal; we imagine BHP, the 'Big Australian', among others, punching above its formidable weight globally. We imagine the benefits trickling down through employment and business opportunity, and through tax dollars. The impression we have is that it's not just the barons who are getting rich from mining either. Coal seems the new get-rich-quick scheme for much of aspirational Australia. Tradies are seizing FIFO opportunities to earn three times as much as they would back home, property investors are cashing in on an unprecedented migration event to coal regions, and a peaceful sleep for superannuants and stock traders seems to depend more and more on the health of China's economy. Economists routinely tell us that our average incomes are greatly inflated thanks to the prosperity generated by the miners. Australians would be forgiven for believing that each of them is a small-time winner in that national game of 'Who wants to be a coal billionaire?' And yet, the impression is an illusion.

As we have seen, the companies behind most coal being mined are overwhelmingly foreign-owned, so most of the money made enriches foreign banks and investment funds, sovereign wealth funds and pension funds behind them, and ultimately the many millions of smaller investors these institutions represent; individuals oblivious to their stake in Australian coal. Meanwhile, most of the wealth generated by the buying and selling of new coal reserves doesn't actually produce coal yet. Until it does very few jobs will be created, let alone tax revenue. Granted, there's an investment boom, but most of the investment is in mines and associated infrastructure, for the near-exclusive use of coal companies. So, while GDP numbers are boosted by the investment flood, and although the boom-inflated Australian dollar increases our purchasing power, it's not really money in our pockets – not for many of us anyway. Most of the money goes to those who have mastered the art of picking undervalued coal reserves that can be on-sold to foreign buyers with very deep pockets. And most of the foreign-owned companies minting Aussie coal barons by seeking to join the free-for-all over the next decade so far do very little actual mining here – Shenhua and Meijin from China; Adani and GVK from India; and many others. Once they start, however, the profits will head offshore to the likes of Messers Adani, Reddy and

Yao, just like most of the profits of current mining.

So, if there's a lesson to be learnt from the rise of Australia's coal barons, it is probably this: Australia's coal boom is not our boom; it belongs to the barons — those with familiar faces and those without any face. And, the sooner we appreciate that, and the dire consequences for the planet, the better.

4 CHARM OFFENSIVE

'It's the pride of Australia's past, and the pride of its future.'
Peabody Energy Chairman GREG BOYCE, talking about Australian coal[1]

The advertisements screened nationwide from early 2011 in the wake of other big ad campaigns by miners aimed at preventing the carbon tax and the Resources Super Profits Tax. Having exercised its muscle so publicly, Big Coal went on a big charm offensive – a series of heart-warming cameos mainly involving mining employees, packaged together under a 'Mining: This is our story' banner. 'We all eat from the same tucker box up here', says a farmer starring in one TV commercial. 'Australians used to ride on the sheep's back. Mate, now they're riding on the back of a dump truck ...'[2] It's a bucolic scene of farmers and miners co-existing happily. 'It's a win-win', says another farmer in the same ad campaign.[3] 'We're all on the same team.' If there's one thing Big Coal wants Australians to believe, that's probably it – we're on the same team – with all the connotations that involves. As we will see, it's a message that the industry hammers home at every opportunity, as if its future depends on us believing it ... which it does.

Myth-building

The 'This is our story' ads are all about building myths, with each one seeking to recast one or more of the many PR problems faced by the industry, from the pitched battle between farmers and miners that is in fact occurring across much of the country due to the impact of mining on water resources, and the plethora of problems associated with the FIFO/DIDO business model. The campaign showcased water recycling in the Hunter Valley, and how the industry pitched in to help save the Central Queensland town of Theodore from flooding.[4] It's much easier to focus on this sort of thing than talk about flooded open-cut mines contaminating rivers across eastern Australia from Central Queensland to the Latrobe Valley; much easier to avoid the issue of long wall mining and how it is triggering land subsidence, causing whole stretches of river to disappear or leak methane gas. The ads give the overwhelmingly male-dominated nature of the coal industry a similarly thorough whitewashing. One clip and series of magazine ads focuses on champion cyclist and coal-miner's daughter Anna Meares, and how integral BHP's sponsorship has been to her inspiring comeback from a bad cycling accident. Another ad features a photogenic mining career mum called Heather who is shown having it all – a senior job at the mine, a tropical lifestyle, a harmonious home life, with ample time left for pony-club instructing, home-renovating and yoga.[5] FIFO blokes are shown similarly contented with their work–life balance, quipping confidently 'happy wives, happy lives'.[6] Reinforcing the industry's interest in a family-friendly image, there's an ad on the Xstrata-funded neonatal intensive care unit at the John Hunter Children's Hospital.[7] The ocker white male miner image gets airbrushed with ads featuring women of French descent, and men of Asian or Islamic backgrounds. There's an Indigenous chemical engineer who doesn't want his 'achievements to be glorified because of my Indigenous background', yet here he is starring in a national ad campaign in which that background (and his mining company employer's support for Indigenous people) is being showcased.[8]

It is all painstakingly designed to positively shape the way we perceive mining's place in the Australian way of life: from the value of its economic contribution to its significance as an employer to its male-dominated culture. Though less soothing, the anti-tax campaigns waged by the coal-dominated mining industry under banners like 'Keep mining strong' also seek to persuade us that coal is a bigger economic contributor and employer than it is in reality, and that the benefits from its expansion are accruing right across Australia. Hundreds of millions of dollars have been spent on such advertising campaigns by the Minerals Council, the Australian Coal Association (ACA), and its members. The full cost of the 'This is our story' campaign has not been fully revealed, but the advertising agency responsible (Lawrence Creative, which also did the 'Kevin '07' and 'Keep mining strong' campaigns) gave some idea of the scale of the operation in its sales pitch about the results. According to Lawrence, '16.92 million people have seen our TV ads on average 8 times ...' Another 8.9 million people saw the magazine ads.[9] The makers of the ads boast proudly, 'Research shows the ads "make you feel good about mining" and "makes me want to go out and get a job with a mine"'.

Conspicuously absent from the list of image problems tackled in the 'This is our story' campaign is the coal industry's biggest problem of all – the greenhouse gas emissions generated when its product is exported and burned in power stations and steel mills around the world. For this, there's a separate multi-million dollar advertising and public relations campaign charm offensive, developed by the same marketing firm. For this challenge, Lawrence Creative was especially creative, rebranding coal as NewGenCoal. A flashy new website (featuring strangely out-of-place wind turbines) sought to reinforce the idea that carbon capture and storage (CCS) is just around the corner.

There were upbeat interviews with scientists from CSIRO, from co-operative research centres and other research institutions, with nary a mention of the extent to which the work of some of these 'independent' experts is funded by the industry. Though the industry puts a tiny fraction of its profits into CCS pilot projects, in combination with slick public relations it's enough to feed public faith that coal may ultimately overcome its emissions problem (see Chapter 6).

The intention is much the same whether it's NewGenCoal or the 'This is our story' campaign or the extensive in-kind support given by mining companies to romanticise the industry through the hit movie *Red Dog*. As it is with the millions more dollars spent by coal mining companies on PR firms – largely to help the industry generate good news stories or 'issue-manage' bad news stories (like coal tankers washed up on city beaches or coal export trains and terminal loaders being blockaded by protesters).[10] On one level it's brand-building through myth-building – taking one of the coal industry's real-world problems and creating the impression that there is no problem; replacing the real story with a story that suits the industry. On another level it's an offer to all Australians to enjoy their own piece of the action, just as those 'ordinary Australians' telling their story in the ads are. Like some fast food commercial designed to make viewers salivate, Big Coal knows that the more Australians find a slice of mining enticing, the less likely they are to mess with the industry's expansion, and that helps to lock in permanence.

It's King Coal's shout

Not all of the coal industry's charm offensive is centrally planned by the industry's lobby groups and PR gurus. Much of it is less about conspicuous PR and more about the benevolent distribution of money to a barely fathomable array of recipients. At times, in the case of the barons who benefit most from the dramatic expansion of coal production, it's as if there is just too much money for so few bulging pockets. Like attention-grabbing volcanoes, we have grown to expect loud and colourful eruptions from our coal barons, followed by a rapid flow of coal dollars in sometimes unexpected directions: such as when Xstrata donated millions to disaster relief after Queensland's big floods, or when Clive Palmer pledged $250 000 to help out bureaucrats who lost jobs in 2012 after cutbacks by the incoming Queensland government.[11] Not all of it is quite so philanthropic, as with Peter Bond's decision to buy the Dunk Island resort devastated during Cyclone Yasi, and even less so with Gina Rinehart's decision to buy large chunks of Fairfax and Channel 10. But whether it's the coal dollars trickling down or the coal barons buying up big, the perception being carefully tended is that the boom booty is being spent here, and that Australian life is being enriched somehow. Life in Australia today, it all suggests, is proudly brought to you by the coal industry.

We get a sense of this in the cities, but the industry's charm offensive doesn't really hit home properly until you visit the coal belt where Big Coal's patronage of life is everywhere you look. Finding something that isn't coal-sponsored is the greater challenge. It doesn't matter where you are — Victoria's Latrobe Valley east of Melbourne, or somewhere along the 1000-kilometre long 'saucepan' of coal seams that roughly semicircle Sydney from Wollongong to Newcastle, with a 'handle' of sorts running north-west to Gunnedah. You could be in any of the dozens of communities in the 40 per cent of Queensland covering the Bowen, Galilee and Surat basins, in which vast coal reserves run all the way down from Collinsville to south of Toowoomba. You could be in the coal port communities of Mackay, Gladstone, Brisbane, Newcastle or Port Kembla. Or you could be on the coal frontier, encroaching further and further into distant, hitherto stranded and untapped reserves, north of the Daintree in Far North Queensland, westwards near Mount Isa, and in Western Australia's north-west. Wherever you are in coal country, the industry's benevolent reach is conspicuously evident. To maximise the PR impact, it's all meticulously documented and packaged – from Xstrata's glossy *What Matters – Corporate Social Involvement* brochures to Rio Tinto Coal Australia's 'Where does our funding go?' online database.[12]

Seemingly, no matter what you want funded, Big Coal is there to help. You want to fund a school camp or an Outward Bound program or spend a year on an overseas exchange or a scholarship to study at university? No problem. How about a preschool learn-to-read program to help with low literacy rates in Cessnock? Great idea! Your Catholic College in Singleton wants to establish an organic garden and chicken pen?[13] Your kindy in Wandoan needs painting inside and out?[14] You want to start a Men's Shed and community garden or run some hip-hop dance workshops or take your show band on tour?[15] Consider it Big Coal's shout. How about some help with your studies to become a hairdresser, a beautician, a model or a diesel fitter?[16] A talented ballerina from the Hunter like you will need dancing shoes, costumes and help with competition fees – have $10 000.[17] Damn it, the babies' playgroup at Nebo will have the swimming instructor it needs.[18] Yes, the Mudgee RSL can upgrade its ANZAC display, and the Wandoan Apex Club shall put in a new barbecue in the local park. Let us all celebrate at Upper Hunter Wine and Food Fair![19] Like a monarch rewarding loyalty from his subjects, King Coal benevolently dispenses favours across his domain. Someone wants to put together a business plan for a gym; someone else is struggling to meet the cost of a funeral. The number of costs that the coal industry will happily cover seems boundless.

On your team?

Nowhere is the benevolence more apparent than on the sporting field. No matter what you or your kids want to play, if you live in a coal community you can reasonably expect the coal industry to chip in to help make it happen. Fancy a coaching clinic from Tennis Queensland? Xstrata is making it available.[20] Maybe your kids would enjoy a coaching clinic with the Newcastle Knights courtesy of the biggest coal railway company in the country, QR National (whose name was recently changed to Aurizon).[21] There's the new 'Xstrata Coal High Schools Rugby Competition', for those kids who prefer another football code.[22] When the Bush Pigs rugby club in Clermont wanted to put a business plan together for its new facilities, Big Coal chipped in. As they did when a talented Aboriginal kid needed help so his family could watch him play in the state Rugby League championships.[23] In the Hunter Valley, the Little Athletics club in Denman needed new uniforms, the Scone Polo Club wanted help funding its trophies, the Aberdeen under 7s soccer team needed a computer, Auskick wanted a sponsor in Muswellbrook, and the Wandoan footy club needed a pie-warmer.[24] Here, a bowls club patio needed an extension; there, a cricket club desperately needed some new gear, and the old urn in their clubhouse needed replacing too. Whatever the sporting wish, coal cash is on hand.

To the rescue?

Often it's Big Coal to the rescue more literally than you might imagine – as the main sponsor for the local helicopter rescue service or the helipad it lands on, or the hospital next door. In 2012, for example, Xstrata chipped in $125 000 to the Central Queensland Helicopter Rescue Service and the Capricorn Helicopter Rescue Service.[25] In NSW the company funds the Westpac Helicopter Rescue Service in the Hunter Valley, as does the NSW Minerals Council, for good measure.[26] Pick a coal community at random and check the local newspaper's daily headlines – say the *Mudgee Guardian* of 27 March 2012 – and you'll find a front page leading with the news that Xstrata has funded the new pathway between the hospital and the helipad.[27] In 2011, when floods struck Queensland, Clive Palmer sent one of his own helicopters to rescue 60 people; staff of Aurizon cashed out $79 000 worth of leave and donated it to the Premier's Disaster Relief Appeal.[28] Having saved the day, the coal-sponsored choppers arrive at coal-sponsored hospitals and health centres. Xstrata bought the Mudgee hospital a new ventilator for the emergency department and the maternity suite.[29] Peabody bought the same hospital a new electronic bed, and they bought new pagers for the local Rural Fire Service.[30] In Gunnedah, having not so long before bought up 43 farms in the region with plans for a large open-cut coal mine in the Liverpool Plains, Chinese government-owned Shenhua wrote a cheque for $1 million to fund the new Gunnedah Rural Health Centre. BHP chimed in with another $500 000 with coal seam gas miner Santos adding $100 000.[31] Whether it's Rio Tinto chipping in to help autistic kids in regional Queensland, or BHP funding video-care for kids with cancer, or QCoal funding the Royal Flying Doctor Service, and Xstrata giving hundreds of thousands of dollars for the Hunter Medical Research Institute for asthma research, the industry is keen to help.[32]

It's no coincidence of course. Community concerns about the health impacts of ever-encroaching coal mines are mounting – from asthma to autism to lead- and mercury-poisoned air and water. Then there are the mental and physical health impacts associated with the incessant infra-noise of mine machinery that carries for many kilometres, not to mention the slow, screeching passage of 100-carriage-long trains that can go on for 30 minutes, and run 24 hours a day. To make matters worse, as the health impacts intensify there is also an influx of new people into the small towns near the new coal projects. With services and infrastructure badly lagging behind, simply getting an appointment with a doctor has become a challenge. An incident in Scone in early 2012 encapsulated the emerging dynamic:

an ambulance rushing an elderly man to hospital with chest pains was forced to wait at the boom-gates for 11 full minutes while a coal train plodded past. He survived the close call, but it's hard not to get the sense that though people may 'eat from the same tucker box up here', the industry eats first, even when your life is hanging in the balance. It's not an impression the coal industry wants to encourage, of course; hence the funding for helicopter rescue. That way they can provide something highly visible that people know can fly them over a jammed railway crossing in an emergency. It's emblematic of much of the coal industry funding – wherever coal is seen as part of the problem, the more assiduously the industry makes itself, or rather its money, an indispensable part of the solution. Health is just one of many issues where this applies.

Friend of tourism?

Tourism is another problematic issue, attracting coal industry attention. In places where coal's increasing presence is viewed as a threat to the local tourism industry, it will often step in to make coal cash part of the answer. So, in Central Queensland, Rio Tinto chipped in $50 000 to the Belyando Council to pay for a Tourism (and museum) Development Officer, with another $20 000 for a Tourism Marketing and Development Strategy.[33] The new visitors centre was to be manned by someone whose position was funded by the nearby Blair Athol Coal mine. The local mayor lauded it as something that would help 'ensure Clermont becomes a "must see" destination for domestic and international tourists' travelling on the nearby highways. In the same region, Rio Tinto gave Mackay Tourism $10 000 for signage, brochures and an internet presence to attract tourists to follow the 'Mining Trail' from Emerald/Blackwater to Collinsville. In the Hunter Valley, it gave the Cessnock Tourist Information Centre $145 000 to increase wine tourism, aided by a TV advertising campaign in Sydney.

Beyond the coal belt, where occasional wrecked coal ships pique public concern about the connection between coal and reef destruction, investments being made in tourism by various coal barons remind the community that there's an upside, such as when Clive Palmer bought up the Hyatt Coolum, or Sam Chong built a new 60-storey hotel in downtown Brisbane to be run by Sheraton, or when Peter Bond bought Dunk Island.

Peaceful co-existence?

It's the same story with agriculture. In communities concerned about the expansion of mining, coal companies are busily funding development strategies for different agricultural sectors, paying for workshops for farmers to improve land management practices, and bankrolling programs to encourage kids into agriculture. The local show is usually sponsored by the local coal company: in Toowoomba and Pittsworth it's Ambre Energy; in Gunnedah it's Whitehaven; in Rylstone it's Centennial; in Mudgee and Gulgong it's Peabody. Same goes for many rodeos. It's all about encouraging the idea that farmers and miners can co-exist peacefully. In regions like the Hunter Valley, for example, wine and gourmet food producers worry about what the ever-encroaching march of coal mines and their toxic dust, noise and pollution means for their wholesome reputation. But Rio Tinto is there, keenly funding various ventures aimed at showcasing local food producers, and even donating tens of thousands of dollars to the Hunter Valley Vineyards Association for market research. Over $26 000 was provided to support the development of a 'Hunter brand' with the idea somewhat wishfully likened to the pristine and pure brands used by New Zealand and Tasmania.[34] In the Upper Hunter Valley, the industry funds a program called Beef Bonanza, aimed at steering local kids into beef cattle and agricultural careers. Up in Central Queensland there's funding in Emerald for a feasibility study on value-added fruit products. When the nation was gripped in drought, BHP Billiton was one of various mining companies enthusiastically sponsoring a rural tour by the Sydney Symphony Orchestra to bring 'soothing Symphony to drought-stricken farmers'.[35] The message they want to get across is, we're all on the same side. In some cases, coal companies are buying up cattle studs, feedlots and even wineries to literally become part of the sector threatened by their expansion.

Going green?

When it comes to the impact of coal mining on the environment, once again the industry is in there funding answers. Growing concern about water quality and availability is just one example. In the Hunter Valley, Rio Tinto is funding irrigation infrastructure for farmers. Up in Chinchilla, apparently as a gift to the community, coal seam gas miner Queensland Gas Corporation has funded pipelines and a wastewater treatment plant, supplying un-potable water from its CSG mining to supplement local agricultural water supplies. This was promoted as 'a solution to the town's water supply problems'. Meanwhile in Central Queensland, Rio Tinto is keenly funding a program to aid the recovery of two vulnerable turtle species on the Fitzroy River catchment. It comes amid growing public concern about government-approved releases of vast volumes of contaminated water in the wake of floods. What impact the more than 7000 megalitres of 'mine-affected' water released by Rio Tinto mines in the 2011–12 wet season had on the vulnerable turtles is anyone's guess.[36]

As concern grows about the impact of mining on biodiversity in Central Queensland, Xstrata is busy funding koala habitat restoration and hairy nose wombat re-introduction programs in the region. As the number of rivers damaged and drained as a result of mining-related land subsidence increases, so too does the industry's funding for catchment restoration programs. On top of this, there's funding for kindergartens that want to run sustainability programs, for councils putting together a sustainability exhibit for the 'Solar Boat' challenge, and sponsorship for the local 'Green Day'.[37]

All about the community?

Then there is the plethora of concerns associated with the FIFO/DIDO employment model that dominates in the coal industry, ranging from the impact of the long absences on workers' relationships with their partners and families to the transformation of coal communities from being places where long-time residents raise families into short-term temporary accommodation for a male-dominated workforce with little stake in the place. There's the extra pressure an influx of miners places on already scarce local services like health, and the increase in traffic and accidents on the roads. There are the many local businesses being forced to the wall by the huge demographic shift under way — as newcomers with everything laid on by the company displace the original customers, and businesses struggle to retain staff lured away to the mines by higher pay or simply forced out of town by rising accommodation costs. To top it off, there's the expansion of economic activity that many coal communities might prefer to do without, from increased alcohol consumption, gambling and prostitution in motels.[38]

For all these FIFO concerns, the coal industry is busy making itself appear to be an integral part of the solution.[39] They fund services to address the problems associated with FIFO lifestyles, from marital breakdown to domestic violence, substance abuse and depression. Xstrata, for example, funds a Dads' Take-Home Reading Program to help build closer relationships between kids and their dads (when they're not away for weeks at a time at the mine), and along with Lifeline and women's refuges, they also fund CityCare Brisbane to help people who are struggling with, among other things, 'dysfunctional home lives' (the sort of lives a FIFO job might trigger). For small- and medium-sized businesses struggling to survive in the new environment in coal communities 'help remains at hand', as Coal & Allied reassuringly puts it, in the form of workshops and a one-on-one advisory service funded through the Hunter Region Business Enterprise Centre.[40] For small businesses across Central Queensland struggling to retain staff amid the coal boom, there are coal industry-funded Better Business Workshops from Commerce Queensland.[41] Whether it's a basalt quarry or a biorefinery, or a company selling indigenous jewellery, the coal industry is in there funding anyone with a project aimed at making communities descending into economic homogeneity look more diverse.[42] In Clermont, meanwhile, as coal's presence crowds out other economic activity, Rio Tinto is funding the local council to prepare a strategy to 'develop a wider economic base'.

For the increasing number of farmers who are weighing up whether or not to stay in farming as coal encroaches ever closer, coal companies are funding planning workshops on 'The future of the family business'.[43] As the influx of mining takes rental prices beyond the reach of long-term residents, coal companies are pledging to build new homes to relieve the pressure. In Clermont, Rio Tinto has gifted land in the town centre to assist the council to provide 25 affordable homes, and is funding local community groups to help 'increase the provision of community services and affordable housing'.[44] In Moranbah, Anglo has recently added 16 new homes with another 28 on the way, while BHP Mitsubishi Alliance (BMA) says it will build over 380 houses in the same region over two years.[45] Clive Palmer's Waratah says it will build 28 new homes in Alpha.[46] In Nebo, Rio Tinto is funding the council to develop an affordable housing strategy.[47]

Traffic carnage is another coal-related problem attracting cash.

The roads in and out of the coal belt have become death traps in recent years, thanks to the explosion of coal-related activity on roads never intended to cope with this sort of traffic. Between heavy transport carrying equipment and supplies to mines and a proliferation of DIDO labour (often involving over-tired drivers), accident rates and the death toll have spiralled on roads like the one between Mackay and Clermont (reportedly 550 crashes in 10 years).[48]

However, rather than cover the cost of making these roads safer, the industry's response is generally to opt for more cost-effective alternatives. So, for example, Rio Tinto is funding driver safety programs. In the Hunter Valley, they give money to the Hunter Economic Development Corporation to prepare its case in support of a government-funded upgrade to the highway from Newcastle's outskirts to Branxton.[49] As the number of people killed in collisions with coal trains rises, Xstrata is funding trauma units at local hospitals.[50] Where increasing rail traffic is turning a cross-town trip into a frustrating commute, as in Scone, the NSW Minerals Council is lobbying governments to build overpasses. Meanwhile, where more and more FIFO miners are causing congestion at regional airports, the industry is paying to be the solution: in Moranbah, BMA is building a new airport. There's seemingly no FIFO-related transport problem for which funding from Big Coal is not sold as the answer.

No strings attached?

For small communities, the incoming tide of funds is truly breathtaking in its scope, and a great deal of it is 'salt of the earth' stuff which no one could possibly quibble about – helping kids play sport, or read, or play musical instruments; planting trees in local parks or installing new barbecues. Who could possibly critique Rio Tinto Coal for giving $10000 to the Cancer Council rather than spending it on Christmas cards?[51] Who can complain about someone receiving help with the costs of a funeral?[52] There's no denying the grassroots nature of most organisations and individuals receiving funding, or the very real benefits involved. Even where the focus is on an issue where the industry is seen as part of the problem, such as health, infrastructure or housing affordability, some undeniable good is being done and generally it comes with no strings attached, beyond the public acknowledgement a sponsor generally expects. Sure, it's standard for coal company grant recipients to participate in some form of positive PR, but this is nothing out of the ordinary. The industry can persuasively argue that its contributions reflect a desire to ameliorate the perceived damage they cause.

However, behind the benevolence, there *are* strings attached. The explicit support comes with an implicit message: that all these worthy things would no longer be funded if the curtain ever came down on coal mining; a great deal more than just the coal industry itself would be at risk if the industry were ever threatened. Life without coal, it is hinted, would be life without a great deal else. Forget the coaching clinics, the new sporting gear, and that trip to the state championships; the future of the show, the music festival, and the new hospital would all be under a cloud. In short, if you want life in Australia to keep rolling on as you're enjoying it now, don't even think about messing with its major sponsor.

This is the last thing on the minds of most communities on coal's new frontier. In many cases, their leaders have watched on sadly as successive generations have abandoned the town for the city, main street shops have been boarded up, rail links, hospitals and schools have withered, and the town contemplates its palliative decline. The sudden and seemingly miraculous tinkle-tinkle of coal is undoubtedly an intoxicating novelty, and many seem to view coal's interest as a first-class, one-way ticket to revival. In Mudgee, Alpha and Boggabri coal sponsorship is a bit like an exciting new drug, its partakers euphoric, if a little hesitant, about where it might all lead. Council staff in Alpha suddenly find themselves in charge of sorting through applications from community groups for grants from GVK and Hancock Coal.⁵³ Other councils like Belyando, Cessnock and Muswellbrook even have Rio Tinto funding staff dedicated to helping members of the public prepare grant applications.⁵⁴ Even so, the dynamic in the more established coal communities is different to the new frontier. In the La Trobe Valley, the Bowen Basin and the Lower Hunter Valley, the novelty and buzz has long since worn off. Surrounded on all sides by coal mines and lunar landscapes, the local economies have been largely hollowed out of industries beyond coal and its ancillaries, and the local businesses servicing them. Here, there's no euphoria of the first drug trip, but the occasionally frazzled dependent resignation of the addict. Coal cash is more like a morphine drip, dulling the reality that there is no way out.

Right across the coal belt the industry's aim is the same, and it's much more calculating and less benevolent than it appears. It's all about fostering addiction to coal-derived funding, and in truth, coal companies, and coal unions for that matter (which are also keen advocates for coal industry expansion and adept at using trinkets to curry favour in coal communities), aren't really buying coaching clinics, barbecues and new urns – they're buying permanence.⁵⁵ When the industry talks about its commitment to sustainable communities, the real focus is on sustaining its own place within those communities; using its financial clout to become indispensable to life in Australia.

Though it's less visible, less ubiquitous and less well appreciated, the same applies at the state and national levels. Away from the coal communities themselves, however, the focus is different. Rather than sponsoring the small-scale stuff, which might involve thousands of kids sporting teams in Brisbane and Sydney, the industry's focus is on conspicuous backing of the more iconic institutions of life in Australia today. From politics to education to the media, arts and sport, coal is in there writing the cheques required to become the major sponsor.

True blue?

Sponsorship of sports is one of the more conspicuous vehicles used by coal companies to ingratiate themselves with Australians. When people see a coal company-sponsored sporting hero, like Anna Meares, for example, odds are they will think of the industry as being just that little bit more true blue. Or maroon, if they happen to be looking at Queensland State of Origin legend Darren Lockyer doing ads on coal-seam gas for Origin Energy. Or red, as in the case of Aurizon's recent sponsorship of the Queensland Reds rugby union team. Or the red and blue of the Newcastle Knights National Rugby League (NRL) team or the Newcastle Jets (the city's soccer franchise in Australia's national soccer competition – the A-League). With the front of the Knights jersey emblazoned with Coal & Allied, the back displaying WesTrac (the coal industry-dominated Caterpillar franchising business owned by the Seven Group), there's no escaping Big Coal's influence on Newcastle's sporting life. The same goes for the Jets. The principal sponsor is Hunter Ports, whose core business is expanding the coal export industry with a new coal loader, and two of the three major sponsors are in the thick of the same coal industry expansion: Aurizon and WesTrac. Meanwhile, in Central Queensland Aurizon has become an official sponsor of Central Queensland's bid for a National Rugby League team. Citing its 4000 employees in the region, the company says proudly, 'As any Queenslander knows, footy is a big part of life for communities right across our state. And it's a part of life that's set to become even bigger.' Thanks, that is, to the cash produced by the coal industry dominating Aurizon's business. The clear message to fans is that they're on the same team as the coal industry, that Central Queensland has no shot at an NRL team without the continued digging, carriage and shipping of coal nearby, and that Newcastle would have no NRL or A-League team without it either.

The tendency of some coal barons to act as saviours for sporting franchises reinforces the message that the future of the team depends on the future of its patron. In 2010, for example, the Knights were on the brink of financial collapse. After years of mismanagement and a $6 million debt, the very real possibility loomed that the rugby-league-mad city would lose its place in the NRL. In walked local coal billionaire Nathan Tinkler offering to buy the club. Tinkler hadn't exactly distinguished himself as a junior rugby league player in Inverell. But when he saved the Knights from collapse he was lauded as a Rugby League hero, even more so when he went on to recruit the game's most successful coach, Wayne Bennett, to join the club. Similarly, when the city's presence in the A-League looked to be threatened in 2010 Tinkler bought the Newcastle Jets; in 2010 he bought a 50 per cent share in Dick Johnson Racing; and he has been one of the largest financiers of thoroughbred horse-breeding in the region (yet another example of coal being perceived as part of the problem – in this case, the encroachment of mines nearer and nearer to thoroughbred studs – positioning itself as a big part of the solution).

Up in Queensland, meanwhile, Clive Palmer beat Tinkler into the A-League, buying Gold Coast United in 2008. As Tinkler no doubt also discovered, this sort of investment buys credibility with the punters and partly inoculates against the tall-poppy syndrome; great PR for the baron and for the coal industry. Ultimately, it went badly wrong for Palmer in early 2012 when a long run of disputes with the Football Federation of Australia (FFA) led to the team's licence being revoked. Yet, here again, the lesson is similar: if things don't go the way of the coal baron funding the sporting franchise your town may lose its team. He may take his bat, or in this case his ball, and more importantly his money, and go home – or as happened here, set up his own association (Football Australia) and launch an 'independent' inquiry into the sport's official governing body.[56]

While Nathan Tinkler's own arguments with the FFA threatened for a time to replicate the Gold Coast experience in Newcastle, today a much bigger threat looms for his sporting empire – the solvency of his own coal business. In less than a year, Tinkler's rise has been savaged by plunging coal prices and the possibility that he may go broke is being taken seriously. A host of sporting enterprises hang in the balance, reinforcing in yet another way that coal cash trickles down to things that the ordinary 'punter' values, but only while coal booms.

Coal-fired education

Educational institutions are another pillar of Australian society increasingly being hooked up to coal industry funding in a range of ways, but most obviously at our universities.

At the University of Queensland, for example, three professorial chairs are funded with naming rights by one of the largest coal mining presences in the state – the BHP Mitsubishi Alliance (BMA). Another Chair in Metallurgical Engineering is funded by Xstrata. The Sustainable Minerals Institute (SMI) at the university is overwhelmingly sustained by money from the mining industry and its industry associations. Of the nearly $70 million in revenue it received over 2010 and 2011, over two-thirds came from mining industry sources, with Rio Tinto, Xstrata, Anglo and BHP Billiton accounting for over $30 million.[57] Much of the work funded does not meet the research definition and instead comes under 'consulting'. In 2009 Peabody Energy, the world's largest privately owned coal miner, paid the University of Queensland an undisclosed sum for eight academics to prepare a document entitled *Coal and the Commonwealth: The Greatness of an Australian Resource*. The 192-page publication, featuring the University of Queensland logo rather than Peabody's, was celebratory and uncritical. 'Coal currently attracts negative press because of the concern that this particular fossil fuel may have on the changing global climate', the report starts off, then adds, 'An often unconsidered reaction ... is to call for the abandonment of coal and other fossil fuels as an energy source. This report shows that this is a totally unrealistic position for Australia and for the world'.[57] And on it went in the same vein. A press release from the university about the report, calling 'continued robust use of coal' crucial, quoted Peabody Energy Chairman Greg Boyce, who said of Australian coal, 'It's the pride of Australia's past, and the pride of its future'.[59]

The agenda being advanced was also highly political. As one of the chapter authors, Marion Diamond, would later reveal, 'The company had planned to use the report in its fight with the Rudd government over its Emissions Trading Scheme'.[60] As it turned out, Peabody's PR people were also very closely involved indeed. In 2012, Rowland PR – was even nominated for a Public Relations Institute of Australia award for what we now find was an extensive campaign around the study.[61] Rowland boasted about the opportunities it seized: to communicate a 'shared commitment' between Peabody and the university to clean coal technologies, for Peabody to be an 'industry leader and advocate in the clean energy debate', and the challenges they overcame – such as combating the impression that the study was industry spin.

With breezy optimism within the SMI, the University of Queensland recently established a 'centre of excellence' for coal seam gas, the main purpose of which seems to be securing the controversial industry's social licence to operate. Of the $16 million raised to run the Centre for Coal Seam Gas (CCSG) in the years ahead, some $15 million has reportedly come from gas companies.[62] The CCSG has even adopted a funding and participation model not that dissimilar to vote-buying, with individual companies allowed to purchase multiple seats on the Strategic Advisory Board (paying $500 000 a year for five years is 'worth one vote').[63] So far, the British Gas-owned Queensland Gas Company has bought four seats with $10 million pledged over five years.[64] As well as providing strategic guidance, these CCSG members also get to 'provide input into which research projects are undertaken with [sic] the CCSG'.[65] The funding provided by industry is untied, and industry can't prevent or insist on a project; but there is no mistaking the influence available via the arrangements.

Elsewhere, the industry has been equally diligent in making itself indispensable for universities, whether it's funding professorships, institutions or lucrative consultancies. At the University of NSW, for example, BHP's coal mining partner, Mitsubishi, has provided $1.1 million to fund a Research Chair in Sustainable Mining, along with a Centre for Sustainable Mining Practices (ACSMP), run out of the university's School of Mining Engineering. Along with a seat on the Advisory Committee, members are told they can enjoy 'being associated with the ACSMP brand'.[66] Here too, the focus is on helping 'mines take the steps needed to maintain their social licence to operate', and on 'the use of the ore body or coal seam, and the optimisation of that resource for humankind'.[67] Along with these industry-funded centres, and the professorships, there are hundreds of scholarships. Rio Tinto Coal Australia, for example, offers mining engineering scholarships at the University of Queensland and the University of NSW.[68] Xstrata offers similar scholarships, plus more at the universities of Newcastle, Sydney and Wollongong.[69] Yancoal also offers scholarships to undergraduates to study engineering at the University of Queensland and the University of NSW. In many cases, the scholarships are not just focused on universities in the coal regions, but targeted at students from there. So, graduates from particular schools in Muswellbrook, Aberdeen and Scone, for example, have a shot of Rio Tinto funding if they want to pursue a mining-related degree. Some of the scholarships are even reserved for company employees. It all serves to reinforce the impression that coal is your meal ticket if you want a bright future. Others are tailored to address negative perceptions – Rio Tinto, for example, funds scholarships at the University of

Newcastle specifically for medicine, nursing and social work – some of the professions that their operations put under increased pressure in coal communities.[70]

Increasingly, Big Coal's reach is extending right into school curriculums, and even kindergartens. In Queensland, for example, 34 'gateway' schools are now part of a coal industry-sponsored collaboration called the Queensland Minerals and Energy Academy. Sponsored by various coal companies and the industry's main lobbying outfit, the Queensland Resources Council, it provides teachers with specially designed teaching materials (dubbed 'Oresome Resources'), including crosswords and fact sheets on coal for children as young as eight, which gloss over the link between coal and climate change.[71] The program is ostensibly 'designed to encourage students to enter careers in the minerals and energy sector', but the other clear aim is to normalise, legitimise, and even glamorise the coal industry for an impressionable young audience, before they form less sympathetic views.

Once they have graduated, there's the promise of a coal-sponsored job or a degree. For those who want to remain in academia, the industry will sponsor that too. Every year, for example, some $16 million is provided to researchers via the Australian Coal Association Research Program (ACARP). Peruse the list of funding recipients and you find hundreds of thousands of dollars at a time being channelled into tertiary research institutions to fund projects that might sustain individual academics (and boost the coffers of their organisations) for many years at a time. Along with the Co-operative Research Centre for Mining (housed at the University of Queensland) and the CSIRO, universities located in coal regions dominate the ACARP recipient list: the University of Queensland, the University of NSW, Monash University, University of Newcastle, University of Wollongong, and the University of Central Queensland. Moreover, you find the same people cropping up again and again, generously funded by the ACA.

And among the list of direct and indirect ACARP beneficiaries you find individuals who wrote the Peabody-funded pro-coal report at the University of Queensland.[72]

Another example is a 'Good practice guide for the Australian coal mining industry', commissioned from the SMI, at the University of Queensland by ACARP (with in-kind support from BHP Billiton, Anglo and Rio Tinto). The study aimed to help the industry manage and minimise cumulative impacts, including greenhouse gas emissions, yet manages to skate past the biggest cumulative impact – the emissions generated when coal is burned.[73] Between the direct funding from coal companies, ACARP and coal industry lobby groups, the industry maintains a small army of academics along Australia's eastern seaboard, whose capacity to raise research funding could hardly be more dependent on the ongoing success of the coal industry. Protests abound that coal industry funding does not compromise academic independence and integrity, and there's no suggestion otherwise here. The work being commissioned through ACARP is the sort of perfectly legitimate R&D you would expect an industry to commission. That said, it's hard to find too many coal-funded academics saying anything publicly that is likely to offend the hand doing so much feeding.

'Coaltural' heritage

Cultural heritage is becoming another attractive target for the coal industry's permeation in Australian life. At the Newcastle Museum you can choose between a Xstrata-sponsored gallery and a BHP Billiton-sponsored display, its slick exhibits carefully designed to leave the visitor with a glowing impression of the significance of coal mining to Australian cultural history as well as to its future. On the coal belt, meanwhile, the industry funds a host of things; Xstrata and Yancoal fund the Henry Lawson Festival.[74] In Clermont, Rio Tinto helped to fund the local museum and more recently they supported an oral and visual history project that culminated in the production of a coffee table book highlighting the importance of mining to the town's history and its future.[75] Then there is the Blackwater International Coal Centre (BICC) sponsored by BMA, Yancoal, Wesfarmers, Jellinbah Resources and Aurizon, among others. It's a museum and open-cut coal mine tour service rolled into one, designed to 'provide a sophisticated and enduring platform for showcasing the mighty Australian Coal Industry and the associated industries that underpin the state and federal economies'. Packaging coal mining as a fun heritage attraction, the BICC is also perhaps as close as it gets to a coal-based theme park.

Beyond this, whatever value the industry places on cultural heritage is belied by their seemingly limitless capacity to fund consultants able to prepare extensive reports that, as if by miraculous coincidence, discover nothing of sufficient cultural worth near proposed new mines that could possibly stand in the way of the project. Whether it's Yancoal replacing 1.7 kilometres of riverbed with man-made channels – destroying Aboriginal corroboree sites, burial grounds, bowls carved into rock and fish traps – or Gloucester Coal pledging to fence off and signpost an Aboriginal honey tree that will literally sit on the edge of an open-cut mine, or New Hope Coal's plans to move the war memorial from Acland, cultural barriers are not allowed to stand in the way of mining.[76] From the graves of children to entire towns, voluminous assessment after assessment finds things of no heritage significance or of local significance only. Never mind the heritage-listed homestead next to the proposed new railway, it's beyond the 30-metre 'impact corridor', the details of significant artefacts can be recorded 'to an archival standard', and, if need be, relocated to an 'appropriate location'.[77] Never mind the old inn or the local cemetery adjacent to the proposed mines; lots of photographs will be taken of everything before the blasting begins just to ensure a digital repository of the local history (see Chapter 2 for more examples), and the blast levels from the mine will be within the legal limits; presumably quiet enough not to disturb the residents forced out by the mine, or to wake the dead supposedly still resting in peace at the nearby cemetery. Never mind the Aboriginal artefacts found on site; they'll either be bundled up and given to the traditional owners for safe-keeping, or, as in case of the 15 000-year-old artefacts found at Rio Tinto's proposed Warkworth mine extension, the company plan is to create a museum to display them.[78]

Doubting the hand that feeds

The overall impact of Big Coal's charm offensive and the relentless flow of sponsorship, beads and trinkets is pervasive, so a community – indeed a country – might see itself as being so dependent on coal industry money that it dare not risk the consequences of changing course. Just occasionally, someone bites the hand doing all the feeding or dares to question it. In Newcastle, a visitor to the coal-sponsored museum recently noted in a feedback survey: 'I know that places like that need sponsors but the industry and coal thing is too dominant, virtually advertising for Xstrata and BHP'.[79] A teachers' union official questions the introduction of coal-sponsored, coal-friendly curriculum in high schools on the NSW Central Coast. It's 'fraught with dangers', he warns; it could well influence the way climate change is taught; we want 'critical citizens not job-ready miners'.[80] When the NSW Minerals Council sponsored a film festival in Dungog, donating staff to run the event, and saturated the program with pro-mining messages, including testimonies from coal-mining employees about their dream jobs, it was too much for some. One filmmaker told the *Sydney Morning Herald*, 'It would be nice if they could just sponsor us instead of trying to brainwash us as well.'[81] According to some visiting students, 'Film festivals are great fun, but this particular one was full of gloss and PR spiel from the Minerals Council about the greatness of the mining industry. Yuk.'[82] Partly in protest, a fringe film festival was held at the same time and featured some films highlighting the damage caused by long wall mining.[83]

In Queensland, a band called The Herd pulled out of the Coal to Coast Festival where it was booked to play when it discovered the event was heavily sponsored by the coal mining industry (indeed, the event was conceived by Dalrymple Bay Coal Terminal staff). As the band stated,

> [we] would never knowingly get involved in an event that supports the coal industry. We believe that alternatives in energy production, as well as employment, should be prioritized as a matter of urgency. Climate change is the biggest crisis facing the planet. Coal is a major polluter. It's pretty simple.[84]

In the Upper Hunter Valley, where coal companies have eagerly funded food and wine festivals,[85] and tried to placate the industry with money for development strategies, a few industry leaders are still willing to speak out in an area that has lost 800 hectares to mining in the past few years. Rosemount's once iconic winery is now a warehouse used by Orica to store explosives used in the nearby coal mines, and the wine region now struggles to accommodate tourists, because it 'is largely booked out by miners and contractors'. As Brett Keeping, President of the Upper Hunter Wine Makers Association and member of the Hunter Valley Wine Industry Association, which received $50 000 for 'research and marketing support' from Rio Tinto, told the *Newcastle Herald*, 'The hard part for us is that it is difficult to attract new investment due to the uncertainty of land use due to the mine expansions.'[86] Bruce Tyrrell is another winemaker happy to speak out: 'A lot of vineyards are being demolished and the coal miners are buying them up … On the one hand we've got the government pushing a sustainability initiative, and on the other we've got vineyards being turned into mining sites.'[87] These objections sometimes make it beyond the local media, but rarely. Almost no one contacted about the sponsorship they receive wanted to say a negative word publicly for fear of the consequences. By and large, silence is bought as community organisations take the trinkets being offered and sign the usual forms requiring them to make positive comments about their sponsors.[88]

Ultimately, none of this is all that surprising. It stands to reason that the coal industry will vigorously serve its interests. All industries do this, though perhaps not to this extreme degree. Nothing in the charm offensive is illegal or illegitimate, and much of it has a very real benefit in the community. What is surprising, however, is how little the industry needs to spend to achieve the desired result. While the appearance is that the industry spends vast sums of money supporting the community, it is a tiny proportion of what the industry makes. Take Xstrata Coal, for example, whose Corporate Social Involvement program in Australia sponsors over 50 different community organisations every year. In all, the company devoted $13.9 million in 2012. That may sound like a lot of money, but it is just over $1/2$ of 1 per cent of the $2.2 billion in operating profit generated over the previous year by Xstrata's coal operations in Australia, and much of it is tax-deductible for the company.[89] A look at other large coal producers, such as Rio Tinto, BHP Billiton and others, uncovers a similar pattern.[90]

The amounts spent on the ubiquitous grassroots sponsorships of Australian life are tiny in comparison to the other sums spent on image management. For example, the 'This is our story' advertising campaign, run to groom coal mythology, cost tens of millions of dollars, as did the advertising campaigns against the Resources Super Profits Tax and the carbon tax. Factor in the unpublished amounts spent on PR firms and it is vastly much more than each of the large coal companies spend on sponsorships and such like. In the end, what makes the coal industry's sponsorship of Australian life different from other industries, and of great concern, is not that it spends relatively more on these things, or that it could afford to spend much more. It's that no other industry in this country has ever threatened to cause harm on the same global scale – not asbestos, nor tobacco, not even uranium. Nor is any other Australian industry knowingly looking to treble its already huge contribution to climate change. None is in the same ballpark or remotely close to it, which is why we should think twice about letting the industry sponsor the ballpark, and so much else.

5 KING COAL'S MUSCLES

'Every day, the focus of my work is to help Australian policy makers avoid caving in to the anti-coal, anti-growth agenda.'
NIKKI WILLIAMS, CEO of the Australian Coal Association[1]

Geoff Graham, a fluoro-safety-shirted, middle-aged man with a slight hint of sweat on his face, appeared glum as he looked out from the newspaper advertisement: 'The new tax on coal mines won't help climate change. So why should I lose my job for it?' Geoff was just one of the 'ordinary' Australians fronting a series of ACA ads opposing the emissions trading scheme proposed in 2009 by Kevin Rudd. Another advertisement claimed that over 7700 jobs would be lost, with little symbols suggesting that the notional jobs to be lost in the overwhelmingly male-dominated coal industry would be equally split between men and women. On the other side of the advertisement the ACA boldly claimed: 'This is the reduction in global greenhouse emissions as a result: Zero.'[2] There was a blizzard of other television, radio and newspaper advertisements, all pointing to a website – CutEmissionsNotJobs.

com.au – which encouraged readers to send an email to members of federal parliament.

The coal industry ads oozed moderation, with one insisting that 'combating climate change is an urgent global issue', and that all they wanted was a 'fair' emissions trading scheme. The subtext, however, was that not one job should be lost or one mine adversely affected as a result of policies aimed at cutting greenhouse gas emissions. Instead, many of them pointed to the coal industry's preferred silver bullet 'solution' – carbon capture and storage (CCS), a prohibitively expensive technology that provides approximately one-quarter less electricity per tonne of coal (see Chapter 6).[3]

The coal industry, one of the most politically powerful players in Australia, was flexing its considerable muscles. Although the industry prefers to work behind the scenes courting government ministers and senior bureaucrats, it occasionally abandons its polite mask to reveal a more domineering side when its interests are threatened. Under successive federal governments for the past 20 years, Big Coal has been a dominant force in undermining and co-opting any move that threatens its power or profits. When the Liberal Party is in power, the coal lobby mobilises the captains of industry as its allies; when Labor is in power it aims to supplement its business network muscle by mobilising mining workers to soften up the Labor Party.

To lend credence to their claims that Rudd's emissions trading scheme would cause massive job losses, the ACA's CEO, former Australian diplomat turned coal lobbyist, Ralph Hillman, pointed to a 29-page study from ACIL Tasman, the coal lobby's economic consultancy of choice.[4] It pleaded for black coal mining to be included in compensation packages for an 'emissions intensive trade exposed' industry. If not, it warned that in the first decade of the scheme 16 mines would close prematurely and there would be 9900 direct and indirect job losses. For those with an eye for detail, the report provided scant verifiable data. At the same time ACIL Tasman was predicting job losses and investment doom and gloom for the association, the ACA posted a link on the front page of its website to a news report stating that 13 new coal mines would soon be opened and $23 billion would be invested in the sector in the next six years. The same article cited the Minerals Council of Australia (MCA), dismissing a carbon tax in Japan as having any impact on imports of Australian coal at the same time that the ACA was claiming a carbon tax in Australia would undermine exports.[5]

The architects of the campaign against Rudd's emissions trading scheme were the Liberal Party's regular election campaign PR firm, Crosby Textor, backed by an advertising campaign designed by former 'Kevin '07' strategist, Neil Lawrence.[6] To outside observers their strategy seemed simple: create a sense of a looming jobs crisis in the coal-mining regions of NSW and Queensland and mobilise the industry's supporters as an army of lobbyists against Labor's regional Members of Parliament. Together with the mining industry's supporters in government and an emboldened Opposition, the industry hoped the public outcry would be enough to entirely kill off Rudd's plan or politically cripple it so that an incoming Liberal government could repeal it. Ultimately, the industry's campaign led to generous compensation being showered on the largest polluters, prompting the Greens, which held the casting vote in the Senate, to vote against the Bill. In late 2009 Rudd's Carbon Pollution Reduction Scheme died an ignominious death and the public standing of the Labor government was seriously eroded.

The defeat of Rudd's 2009 carbon emissions trading scheme had a profound effect on the peak mining industry groups. Their coffers swollen by the proceeds of special levies, the key lobby groups were in a belligerent mood. Overlapping the debate of Rudd's emissions trading scheme was an overhaul of the entire Australian tax system.

In its submission to the review, chaired by then Treasury Secretary Ken Henry, the MCA cautiously urged that state-based mineral volume-based royalties be changed to a profits-based system, but only for new projects.[7] The review team's final report broadly followed the mining industry's suggestion, but before long the industry grew alarmed that it was being frozen out from finalising the all-important detail of the changes. Mitch Hooke, CEO of the MCA, warned his board and began developing campaign plans on the assumption that 'we're going to be going to war'.[8]

The Resources Super Profits Tax was to become the centrepiece of the final tax package announced by Rudd and Treasurer Wayne Swan with a detailed revenue estimate and the headline tax rate of 40 per cent on all minerals announced shortly afterwards in the federal budget. Within weeks the industry launched what was to become a six-week onslaught of 'Keep mining strong' advertisements railing against the new tax. The ads had a devastating effect, leaving Rudd increasingly isolated and Labor's support greatly diminished. A month into the blitz, Keith De Lacy, a former Queensland Labor Treasurer who became Chair of Macarthur Coal a few years after leaving parliament, called for Rudd to be dumped.[9] With a federal election looming, the clamour against Rudd resonated with Labor powerbrokers. Less than two weeks after De Lacy's call, Julia Gillard launched a leadership challenge and Rudd resigned. From the outset, Gillard signalled that negotiating a peace deal with the mining industry was her highest priority. For its part, the MCA suspended its ads while an in-principle deal was hammered out between senior Gillard government ministers and representatives from Rio Tinto, BHP Billiton and Xstrata. Treasury officers were pointedly excluded from the negotiations.[10]

Ultimately, the new Mineral Resources Rent Tax would apply only to coal and iron ore, but there were huge loopholes. Instead of the federal Treasury reaping $10 billion a year, the yield was savagely pared back to, at best, a few hundred million dollars in the first year.[11] Rio Tinto for one, paid nothing in the first year and may never.[12] All up, the campaign to overturn the proposed tax cost the MCA just over $17.2 million,[13] small change for an industry that pays more for a handful of ore haul trucks.

After the Gillard government barely held onto power at the August 2010 election it introduced a carbon tax, a price of establishing a minority government with the Greens and Independents. Once more, the coal industry targeted coal regions in Queensland and NSW with a campaign full of dire predictions from yet another ACIL Tasman report, this time claiming that by 2015 eight existing coal mines would close early, with a further 18 coal mines to close by 2020. All up, they claimed that 5000 jobs were at risk, and over 12000 jobs across the economy.[14] Citing the same study, the head of the Queensland Resources Council extrapolated further, 'assuming a four times indirect employment multiplier, this means the loss of 13000 Queensland jobs'.[15]

As always, the devil was in the detail. Few people would have realised from these warnings that, according to the federal government, the fall in the export price of coking coal was almost 60 times greater than the average impact of the carbon tax on open-cut coal mines in Queensland.[16] Data buried in the ACIL Tasman report indicated that at worst, even with the carbon price, the coal industry was likely to grow by 25 per cent in a decade, just not quite as much as it could have.[17] A journalist at the *Australian Financial Review* scornfully pointed out that the ACA's claims of 'job losses' were not of existing jobs, but of jobs that might have been created if there had been no carbon tax. It was 'not at all the same as putting that many real people out of work', the journalist pointed out.[18] A few hard-headed investment analysts commented that the carbon price 'was unlikely to force significant mine closures'.[19] The veteran business journalist, Ian Verrender, commented that 'the idea that an industry can hoodwink an entire nation by spending some small change on glitzy television advertisements never ceases to amaze'.[20]

The country's dirtiest coal power generators in Victoria's Latrobe Valley had an even bigger card to play: the spectre of blackouts. Energy Australia's managing director, Richard McIndoe, pointed to the prospect of a carbon tax and warned, 'I think there's going to be real challenges during the peak times'.[21] It was a theme the industry hammered relentlessly.[22] The scare campaigns had their effect. To mollify industry, the government included a major loophole, provided generous compensation to mining companies, and pledged to pay $5.5 billion in cash and free carbon permits from an 'energy security' fund to the most polluting power generators so they could keep on polluting.[23]

After the best part of two decades, where the mining industry had grown accustomed to setting out its agenda publicly and following up behind the scenes to 'work things through' with federal governments of both major political persuasions,[24] the game had changed. In just three years the mining industry had waged and largely won three bitter fights with the federal Labor government where they relied on big-budget advertising blitzes to turbocharge their behind-the-scenes lobbying campaigns. The visible display of the industry's raw power has served to enhance its influence in the minds of key decision makers, ensuring subsequent campaigns would be shorter, sharper and cheaper. Ahead of the 2012 federal budget, the government flagged that it was considering axing the mining industry's generous diesel fuel rebate. Once more, the MCA launched an advertising campaign, and the next day the government capitulated.[25] After revelations in early 2013 of the paltry tax take of the Mineral Resources Rent Tax, some Labor Party ministers pondered tweaks to yield more revenue. Once more the MCA launched another round of ads aimed at the government.[26] A bruised government took the hint and fell quiet. A lack of public debate over policies that threaten their interests is just what the coal lobby wants.

In just a few years a handful of mining industry groups could validly claim to have killed off the emissions trading scheme, toppled a prime minister, gutted the extension of the Resources Super Profits Tax, weakened the carbon tax to the point where it would have minimal impact on domestic consumption and next to none on exports, and preserved the diesel fuel rebate subsidy. For the cost of a little over $100 million on tax-deductible lobbying and advertising on these four campaigns the mining industry has saved itself tens of billions of dollars, paved the way for its expansion and locked in unquestioning support from the Liberal and National Party Opposition.

Following the 2010 ousting of Rudd and the crippling of the Resources Super Profits Tax, Rio Tinto's CEO, Tom Albanese told 500 mining executives in London that other governments around the world who wanted 'a new tax to plug a revenue gap' must 'learn a lesson' from the Australian events.[27] *The Australian* was in no doubt about the meaning of this statement, headlining the speech as 'A warning to other governments'.[28]

Beyond the drama of the political skirmishes, the mining industry has effectively created a veto over government policy for all but the bravest of governments.

The greenhouse mafia

To outsiders, the power of the fossil fuel interests to shape Australia's policies on climate change may seem new. But for the last two decades the coal industry and its allies have had the upper hand on just about all the key elements of Australian government climate policy. Keeping Australia's political agenda coal-friendly is no accident and has spawned a boutique industry of lobby groups, which, on conservative estimates, now spend over $40 million a year and have dozens of staff devoted to serving Big Coal's political interests. At its core are five lobby groups hunkered down in Canberra, Sydney and Brisbane with tentacles stretching into all the key companies and industry associations, ministers' offices, government agencies and industry-friendly economic consultancies.

The leader of the pack is the MCA – dominated by companies such as BHP Billiton, Rio Tinto, Xstrata and Peabody Energy – with a budget of over $32 million[29] and over 30 staff. Its political partners on coal issues are the ACA, the NSW Minerals Council and the Queensland Resources Council, each now boasting income of over $9 million a year. While the final member of Big Coal's core team, the Australian Industry Greenhouse Network (AIGN), only has income of just over $470 000 a year and barely two full-time staff, its size belies its influence. There are also other lobby groups for the coal-based power generators, such as the Energy Supply Association of Australia (ESAA), along with the state governments, some of which still own coal-fired power stations.

The key to wielding invisible power for these lobby groups is gaining access by hiring former insiders with deep personal connections to the main decision makers: ministers, ministerial advisers, key government departments and key media players. If a company needs to discuss a delicate topic with a minister or influential public servant, a former insider will know if an invitation to a corporate box at the football or opera is the way to go or what points an economic consultant's report needs to hit with pinpoint precision.

The coal lobby hasn't always been so formidable. When Australia took its first steps to grapple with the issue of greenhouse gas emissions in the early 1990s, during Bob Hawke's prime ministership, the Australian Mining Industry Council (later to become the MCA) and the Business Council of Australia (BCA) responded haphazardly at first. As it became apparent to industry that greenhouse policy would be a permanent fixture on the political agenda, AIGN was created as an informal industry umbrella group sited in the MCA's office. AIGN has long been the greenhouse policy powerhouse of industry groups, where policy differences between the dozen industry associations and 23 corporations it represents are ironed out so that a united front can be presented to government. While AIGN's membership appears broad, the power has always been held by the coal miners, power generators and the broader mining industry. AIGN's clout is reflected in it having been included as the industry representative on the official Australian delegation to the UN climate negotiations for over 15 years.[30] Under a cloak of anonymity, members of the network – insiders who have spun through the revolving door from government to work for industry – have admitted to leveraging their personal connections to help draft cabinet submissions, briefing notes and minutes on government greenhouse policy.[31]

At the outset though, AIGN was a new kid on the block at a time when industry was reeling in the face of a wave of public attention on environmental issues. 'Ecologically sustainable development' was all the rage and industry groups were struggling to regain control of the public agenda. Even though public and media coverage of environmental issues blossomed, the coal industry was still thriving: electricity demand was constantly climbing, new coal power stations were being built and coal exports were proceeding with little opposition. The main game for industry was trying to ensure that the federal government didn't agree to anything in international climate negotiations that would drive domestic policy. The most powerful way to shape the thinking of the Keating government was by buying backdoor influence via the reports of the government's own economic modeller, the Australian Bureau of Agricultural and Resource Economics (ABARE). ABARE enthusiastically threw open its doors to industry, as it had a target of raising 30 per cent of its income from external sources. For $50 000 a year, industry groups active in AIGN – including the ACA, BCA, the Electricity Supply Association of Australia, BHP and CRA, a precursor of Rio Tinto – all got a seat on a steering committee created to design and oversee a project to model the impact of adopting greenhouse gas emissions reduction targets.[32] The dire claim subsequently made by ABARE, that the economy would take a hit of $150 billion and result in the loss of tens of thousands of jobs, was a powerful lobbying tool designed to soften public concern and provide cover for industry-inspired policy retreats during John Howard's decade-long reign as prime minister.[33]

Ahead of the pivotal 1997 Kyoto climate conference the Howard government bowed to domestic and international pressure and announced its intent to establish the Mandatory Renewable Energy Target. The scheme, which came into effect in 2000, required power generators to supply 2 per cent of their generation from new renewable energy sources and was aimed at accelerating the growth of the wind and solar industries. For the coal industry, 2 per cent allocated to renewables in an electricity market that was already expanding by that or more a year would have next to no impact. For the Howard government, it was convenient window-dressing designed to mask Australia's extraordinary ambit claim of wanting to win international approval for increasing greenhouse gas emissions at a conference aimed at negotiating cuts.

To the initial delight of the coal and other industry groups, Australia's delegation to the Kyoto conference succeeded in winning approval for an 8 per cent increase in greenhouse gas emission levels over the 1990 baseline. A few hundred thousand strategically directed dollars for an ABARE study, and a bit more on lobbying, had bought Australia's biggest polluters the international community's reluctant approval to increase emissions for another 15 years. However, once the initial euphoria faded, the key players in AIGN decided the Renewable Energy Target was no longer tolerable. In the course of protracted negotiations with the states, prior to its final introduction in 2001, one industry lobbyist went as far as writing a ministerial brief while working from within a government department arguing against the costs of the scheme.

When this was later made public, the head of AIGN, Robyn Bain, defended the industry's dual role on the grounds that it was 'important information that the government needs'.[34]

Just a few years after the scheme commenced, Howard and his Minister for Industry, Tourism and Resources, Ian Macfarlane, were complaining that it had been too successful – especially in stimulating investment in wind farms – and convened with the Lower Emissions Technology Advisory Group (LETAG), a government-created advisory group that included key AIGN members, including BHP Billiton and Rio Tinto. Their role was simple – to provide advice to the Howard government on the proposed content of what was to become his 'Securing Australia's Energy Future' policy. The wind industry wasn't invited.[35] However, leaked minutes of a LETAG meeting with Howard and Macfarlane revealed the government had a problem: if they were going to kill off the publicly popular Renewable Energy Target, they needed to be in favour of something else, preferably a joint industry and government technology fund. For its part, industry representatives were receptive to a government scheme which, as one put it, wasn't 'skewed to wind power'. Macfarlane and the coal and power industry representatives knew they could all agree on CCS, which aimed to capture carbon dioxide emissions from power stations. Other company executives wanted a fund open to other approaches. Despite Howard's intention to abolish the Mandatory Renewable Energy Target (MRET), a review of the target initiated by Howard and undertaken by former Northern Territory Senator, Grant Tambling, proposed the scheme be expanded. Howard took the safe option and left the scheme as it was, but embraced the coal industry's option of CCS as well.[36] By late 2006 Howard's increasing public persona as a climate change sceptic and opponent of renewables was increasingly out of sync with the public mood.

A week after the Labor Party elected Rudd as Opposition Leader, Howard changed tack once more, announcing that he was again open to the idea of an emissions trading scheme. Supporting such a scheme could create the public impression that he was serious about taking action on climate change and win the support of business. The membership of the Prime Ministerial Task Group on Emissions Trading made Howard's inclination clear: on it were representatives from BHP Billiton, Xstrata and International Power. In a media release announcing the formation of the group, Howard described the 'possession of large reserves of fossil fuels' as one of Australia's 'major competitive advantages' that 'must be preserved'. One solution he flagged was the 'the development and deployment of low emissions and clean coal technologies'.[37] Within the Department of Prime Minister and Cabinet a 'joint government–business' secretariat was established to prepare the task force report. Along with senior public servants from a range of government agencies were two staff seconded from industry lobby groups: John Daley from AIGN and Maria Tarrant from the BCA.[38]

The final report was predictable: earnestly flagging that something needed to be done domestically about global warming, but only in a way that limited costs. On one hand there was opposition to the government for 'picking winners' by supporting renewable energy, but on the other, there was an insistence that companies such as coal-fired power generators be given free greenhouse gas emissions pollution permits to 'compensate firms for a disproportionate loss in asset value'.[39] However, before he could implement the recommendations, Howard was swept from office by the Labor Party, headed by Rudd, who quickly ratified the Kyoto Protocol and in 2009 expanded the Renewable Energy Target to 20 per cent by 2020.

While the 2007 change of government altered some significant elements of Australian climate policy, others remained the same under both the Rudd and Gillard Labor governments. Australia's timid emissions reduction target – set at a 5 per cent reduction by 2020 from a 2000 baseline – is one endorsed by the AIGN. In an echo of Howard-era rhetoric, deeper cuts will only be embraced if the largest emitters and other developing countries are part of a binding international agreement, and there is little chance of that occurring any time soon. Aside from implementing a carbon tax out of political necessity, little else has changed for the better during the Gillard government. Indeed, as its political stocks sank in early 2013, it embraced the Liberal Opposition's policy of abolishing the Department of Climate Change.

The inside lane

As a controversial boom-time industry facing increasing opposition, coal companies have turned to former political insiders to steer them through tricky times. Recruited from Labor, Liberal and National Party ranks, the coal industry provides a lucrative career path to former ministers, advisers and bureaucrats. Whether as corporate executives or indirectly as company lobbyists and consultants, this small army of coal boosters helps to lock in both state and federal government and Opposition support for the industry, regardless of who is in office. Much of the attraction for companies and lobbying firms is that former insiders can detail what buttons to push to make sure ministers approve their projects.

One of the reasons government climate policies change so little is that no matter who is in government, the power of the industry ensures that ministers have an open-door policy for the coal industry, and that, more often than not, the people who visit have previously worked with the minister, his advisers or the government. When governments change, one set of doors for former ministers and their advisers closes while another opens.

These days, former Deputy Prime Minister from the Howard government, Mark Vaile, chairs the board of Whitehaven Coal,[40] and is estimated to have a stake in the company worth over $10 million.[41] Asked recently if he would ever return to politics, Vaile laughed stating he didn't miss it. 'And, no, I can't afford to go back there'.[42] Howard's former chief of staff of nine years standing, Arthur Sinodinos, was, until his recent move into the Senate, Chair of Blackwood Coal (now Cuesta Coal), a coal exploration company with interests in Queensland. In his last three months with Blackwood, Sinodinos earned just over $70 000 in fees and shares.[43] Guildford Coal, a small coal company hoping to develop a project in Queensland's Galilee Basin neatly illustrates the bipartisan revolving door dynamic. Former Parliamentary Secretary in the Howard government, Peter Lindsay, is Chair of that company, while another of its directors is Alan Griffiths, a former federal Labor Resources Minister. Guildford Coal's general manager of stakeholder relations is Tony Mooney, a former Labor mayor of Townsville. All of these engagements are paid, and they not only enable companies to gain insights into government decision-making, but engender good relations with at least one side of politics or the other.

Australia's longest-serving Treasurer and former Deputy Leader of the Liberal Party, Peter Costello, is just one former politician who has joined a lobbying firm boasting coal companies on its client list. Costello is Chair of the Advisory Board of ECG Advisory Solutions, a boutique Victorian lobbying firm founded by two of his former political staffers, Jonathan Epstein and David Gazard.[44] Among its rollcall of corporate clients is Exergen, a small Victorian brown coal company which will live or die on whether it can win a new brown coal allocation from the Victorian Liberal government.[45] The company is also aiming to benefit from a grant of up to $40 million from the federal and state governments' new $90 million Advanced Lignite Demonstration Program.[46]

Until late 2011 former Foreign Affairs Minister, Alexander Downer was on the board of Clive Palmer's Resourcehouse Ltd, and has joined the board of one of Gina Rinehart's companies.[47] Downer is a partner, along with former Labor minister Nick Bolkus, in the lobbying firm Bespoke Approach, which has as clients Xstrata and the Noble Group, a commodities company with significant coal interests.[48] Downer says that while he doesn't lobby politicians directly, he has advised a mining client on the mining tax. 'In terms of domestic politics, we give them advice on how they should approach the government. I think we can give very sound advice because we know all these people, either Liberal or Labor,' he explained.

As an active player in Liberal politics, Downer can easily keep his finger on the Liberal pulse: 'I am a vice president of the Liberal Party and I also run fund-raising in South Australia. I host a lot of business lunches and dinners for the Liberal Party, so if people want to know what the Liberal Party thinks about XYZ, I can tell them.'[49] Former NSW Premier Nick Greiner was hired by a consortium of coal companies to lead their lobbying campaign ahead of the privatisation of Queensland Rail.[50] Former Labor Queensland Deputy Premier Jim Elder is the owner and executive director of lobbying firm Enhance Corporate, which has among its clients two of Clive Palmer's companies: Queensland Nickel and China First, which is seeking to develop the 40-million-tonne a year coal project aimed at the Chinese market.[51]

Even the coal industry lobby groups hire lobbying firms to help them. The Queensland Resources Council has hired David Moore as a lobbying consultant. Until 2007 Moore worked as chief of staff for Mal Brough, a federal minister in the Howard government, before he established a lobbying firm, The Next Level Consulting Service. For a year before the 2012 state election Moore had a stint as the Chief of Staff for then Opposition Leader Jeff Seeney. Seeney is now the Deputy Premier and Minister for State Development, Infrastructure and Planning, and a key player in decisions on new coal and coal-seam gas projects. After the election of the Liberal–National Party, Moore returned to his lobbying business and his client list swelled. These days his clients include Bandanna Energy, GVK Hancock, Metrocoal, Queensland Coal Investments and Stanmore Coal.[52]

While lobbying is perfectly legal, the growing army of coal industry lobbyists reflects the sheer volume of new mines that are demanding decisions by government ministers and agencies. In a review of lobbying disclosure standards, the NSW Independent Commission Against Corruption warned in 2010 that the corruption risk associated with lobbying 'is that those who are powerful or wealthy enough to understand how government works or to engage the services of someone who can navigate the decision-making maze on their behalf will exploit their position to their advantage and to the detriment of the public interest'. It also warned there was a risk that lobbyists 'may build up relationships with public officials or exploit existing relationships in order to exert improper influence'.[53] Even if no coal lobbyist ever used improper influence, the problem remains that the cumulative effect of dozens of well-connected lobbyists all arguing for more coal mining ensures that any dissenting views struggle to get a hearing.

While courting decision makers is one side of Big Coal's lobbying coin, they are not above trying to silence dissenting points of view either. Ahead of the 2007 NSW election, the NSW Minerals Council launched a newspaper, television, billboard and web-based advertising campaign in the Hunter Valley promoting the benefits of mining. One image featured a surfer out on the ocean with the tagline, 'A cleaner future: brought to you by mining'. It was too much for the coal activist group, Rising Tide, which launched its own website with spoof anti-ads. In their version, over the top of an aerial photo of a huge open-cut coal mine were the words 'Destruction: brought to you by mining'.[54]

The NSW Minerals Council CEO, Nikki Williams, opted for a legal sledgehammer. Lawyers working for the NSW Minerals Council, Freehills, complained to Rising Tide's internet service provider that the counter ads were a breach of copyright and had the website closed down. After re-hosting the website overseas and even tweaking the ads, the Council kept up its legal threats, but after a scornful public reaction, the legal threats quietly faded away. While the coal lobby may be keen to use the law to stifle dissent, they don't like the idea of community groups being able to take legal action against coal companies. Shortly after publicity in March 2012 over the NSW Environmental Defenders Office (EDO) being on hand to provide legal advice at an anti-coal meeting, Williams, with her new ACA hat on, wrote to NSW Premier Barry O'Farrell urging that any public funding of the EDO be restricted. Of particular concern, she wrote, was that the EDO had run three cases against decisions on coal projects.[55] Eight months later, the NSW Mineral Council's CEO, Stephen Galilee, went further, urging that all legal aid funding for the group be scrapped 'as a matter of urgency'.[56] O'Farrell responded by cutting funding for the EDO and forbidding state agencies to provide funding to groups 'providing legal advice to activists and lobby groups'.[57] Academics also attract the ire of Big Coal's lobbyists. When a researcher at Queensland University of Technology publicised a couple of papers highlighting some of the social and economic downsides of FIFO mining operations, the Queensland Resources Council's Chief Executive, Michael Roche, wrote to the Vice-Chancellor of the university, calling it 'leftist, anti-mining company advocacy' and 'biased in so many ways'.[58]

A study for every season

The public ammunition of choice for coal lobbyists is invariably an economic study: one that predicts boom-time joy if a preferred policy is delivered, or dire economic consequences if it isn't. Few doubt that a pile of economic studies building on the 'coal is great' narrative will have a powerful effect on framing a debate. For most politicians, a seemingly authoritative economic study provides the prepackaged quotable quotes they want for a media doorstop interview or to cite in parliamentary debates. For sympathetic public servants or advisers, a study provides an analytical framework to be adopted or, at worst, partly accommodated. For any doubter, it is too risky to entirely ignore an economic study from a powerful lobby group. Few politicians or public servants, even if uncertain about the merits of a study, have the skills or time to unpick the often outlandish assumptions studies often seem to hinge upon. As a result, most studies go undisputed.

In the absence of any alternative views, the din generated from coal lobby groups and their consultants' reports reverberates around the media echo-chamber that surrounds the decision makers. The cumulative impact of study after study – one here by the coal miners and another there by the power generators – builds a narrative that few dare ignore. The effect is reinforced by the absence of countervailing studies from the industry sectors that stand to lose most from the coal boom's downsides. For example, farming lobby groups, normally vociferous advocates on often obscure policy issues, have remained largely mute about the role of ever-increasing coal exports and consumption and the impact of more extreme weather events from human-induced climate change.

The power of the coal industry's studies is often amplified by uncritical single-source reporting by some journalists who are unwilling or too pressed for time to test key assumptions. Often these consultants' reports are delivered to hand-picked business journalists as an 'exclusive' on the unwritten assumption that it will be given prominent and significant coverage. It is rare for such exclusives to feature much in the way of dissenting views or, if there are any they are buried away at the back of a long story. For example, The Australian's coverage of ACIL Tasman's June 2011 report for the ACA on carbon tax was billed as an 'exclusive' report on the 'independent' modelling, but quoted no dissenting points of view, not even from the government.[59]

The point of such reports is not only to garner sympathetic media coverage, but to use them as a delivery mechanism for a lobbying demand. In the case of ACIL Tasman's report, the ACA said it wanted just two changes to the carbon tax scheme: a phase where auctioning of permits would be set at a sufficiently low level that industry's competitiveness would not be affected and an exemption from the carbon tax on methane emissions until international competitors introduced one. Other than citing the author of the ACIL Tasman report and ACA's director, The Australian's article cited only one other source: the National Secretary of the Construction Forestry Mining and Energy Union, Tony Maher, who was demanding compensation for 'gassy mines'. The government relented and, when the final package was announced it included $1.3 billion in compensation for mines with high methane emissions and a further $70 million for a fund to research new technology to reduce methane emissions.[60]

The power of one

One person who favourably cited the ACIL Tasman report as evidence that a carbon tax would have terrible consequences for the Australian economy was Australia's richest woman and WA mining billionaire, Gina Rinehart.[61] Rinehart's grim opinion of a carbon tax and quirky opinions on climate science are hardly rare in the mining industry, but matter immensely because of her massive media investments. In November 2010 she invested $166 million for a 10 per cent stake in Network Ten, which operates three television channels across five capital cities. The following month, she became a member of the board of directors.[62]

On the surface, the media investments may appear to be just another hobby or indulgence by a wealthy person looking to broaden her interests and perhaps attract recognition. Yet, this seems unlikely, given that Rinehart's Hancock Coal has billions of dollars at stake in Queensland and the future of her investment hinges on public sentiment about the issue of climate change in general, and on the legitimacy of Australia continuing to expand coal exports, regardless of the carbon dioxide emissions generated when it is burned offshore. Thus, Hancock has an enormous vested interest in how the media covers climate change in general and coal exports in particular.

At the time of her investment in Fairfax, Rinehart's company made it clear in a media statement that she was 'interested in making an investment towards the media business, given its importance to the nation's future'.[63] Climate change sceptic and *Herald Sun* columnist and blogger, Andrew Bolt, wrote of his 'strong and not entirely uninformed hunch' that Rinehart's motivation was about issues 'involving our country's future and threats to the wealth we've taken for granted'. Bolt coyly wrote that 'I can't disclose just why I suspect that', but, after pointing to a Hancock Prospecting media release, concluded that 'Rinehart is on a mission. Channel 10 is just the vehicle'.[64] He was right. At her very first directors' meeting in January 2011 Rinehart reportedly asked senior executives to consider having Andrew Bolt on the network.[65] A few months later it was announced that Bolt would host a Sunday morning television program, 'The Bolt Report'.[66]

By then, Rinehart had also started acquiring shares in Fairfax Media. At first, in late 2010, she quietly spent approximately $100 million for a 4 per cent stake and sat on it. A little over seven months later climate change sceptic Christopher Monckton, who was touring the country with support from Rinehart, spoke at the Mannakal Economic Education Foundation, a Perth-based right-wing think tank, also supported by Rinehart.[67] At the event Monckton bluntly stated that, 'whatever you do at a street level ... is not going to have much of an impact compared with capturing an entire news media'. Monckton proposed that a group of those present develop a business plan for a news channel, which could feature people like Andrew Bolt and climate change sceptic and blogger, Jo Nova. Monckton wanted to take the plan to some 'super rich' people who could be 'angel' investors in the new media venture promoting libertarian, free market views.[68] Whether by sheer coincidence or not, six months later Rinehart spent another $192 million for a 9.9 per cent stake to become the largest investor in Fairfax Media.[69]

As investments, her stakes in Network Ten and Fairfax have been failures, but as political advocacy vehicles they are invaluable. The potential for Rinehart to shape the way news is reported is large given the relatively small investment, and it extends way beyond Network Ten. Rinehart's shareholding in Fairfax not only involves the country's pre-eminent financial daily newspaper (*Australian Financial Review*), but also the two largest and most open-minded national newspapers (*Sydney Morning Herald* and Melbourne's *Age*). It also extends into the coal belt itself to various newspapers owned by Fairfax — *The Land, Newcastle Herald, Illawarra Mercury, Lithgow Mercury, Dungog Chronicle, Mudgee Guardian, Scone Advocate, Maitland Mercury, Muswellbrook Chronicle, North Queensland Register, Latrobe Valley Express, Collie Mail* in Western Australia and many more. Added to that is a swag of influential metropolitan radio stations.[70]

Rinehart's mere presence as a shareholder, with directors representing her interests, is likely to have the desired effect. When seeking a seat on the board of directors of Fairfax, she refused to agree to a condition that she abide by Fairfax's charter of editorial independence. Rinehart's friend, John Singleton, leapt to her defence and was adamant that she should have the right to hire and fire editors and senior reporters. 'I think any business has the right to do that,' he said, if the company was being 'held to ransom by your employees', especially if it meant the company was going to lose money.[71]

And Rinheart and her company are not subtle when it comes to dealing with journalists. Shortly after her media investments became public, she declined an interview with *Forbes'* journalist Tim Treadgold, although she indicated that she might reconsider his request at a later date after reviewing his published article to determine whether it was 'anti-mining'.[72] When her company was organising a media event at a proposed Queensland coal project, journalists were asked to sign a waiver, stating that:

> there should not be reporting of other events which may inadvertently transpire on the day regardless of how newsworthy you may consider them — e.g. critical incident, unplanned or unscheduled event, or any information on other matters not related to the project or related to the project should they be overheard in conversation.[73]

Rinehart's company, Hancock Prospecting, also had a subpoena issued against three journalists demanding that they give her records from any contact with any of her three children, who are locked in a bitter legal dispute over a family trust.[74]

From cautiously expressing her concern about 'media sensationalism' driving a political response on climate change,[75] Rinehart has become more adamant that global warming isn't real,[76] claiming that carbon dioxide is mere 'plant food'[77] and railing against the media largely reporting the views of 'climate extremists'.[78] As she has become more outspoken on climate change, she has also become more enamoured of supporting climate change sceptics. She has appointed one of the most prominent sceptics, geologist Ian Plimer, to the boards of three Hancock Prospecting companies,[79] and funded his visit to Perth in 2011 to address the political leaders and advisers at the Boao Forum for Asia Perth conference.[80] When asked how much Rinehart donated to 'organisations and individuals that are challenging the mainstream climate scientists' there was no response.[81]

Crisis, what climate crisis?

Gina Rinehart's support for the climate change sceptics is just the tip of the iceberg. Leading coal mining companies such as BHP Billiton and Rio Tinto,[82] as well as a dozen or more of Australia's dirtiest power generating companies, have long supported the Melbourne-based Institute of Public Affairs (IPA), an organisation largely unknown to most Australians. While it purports to be merely a 'think tank', for almost two decades it has been central to promoting the interests of the mining sector and its campaign promoting doubt about climate change. It has disputed the significance and costs of likely climate change, defended coal mining and coal-fired power stations, and opposed both the carbon tax and Rudd's proposed emissions trading scheme.

Mike Nahan, the current Minister for Energy and Finance in the WA government, explained to a corporate PR conference in 1996, when he was Executive Director of the IPA, that the best way to counter Greenpeace was to 'develop a countervailing ratbag organisation. If you have left wing organisations and right wing organisations one tends to cancel out the others ... you have to assist fund a variety of individuals, such as universities, and a variety of organisations'.[83] The IPA's current Executive Director, John Roskam – a former Rio Tinto PR executive – has boasted about the group having 'helped and supported just about all' of the serious sceptics in Australia 'one way or another'.[84] Roskam listed Plimer, retired oceanographer and IPA Fellow Bob Carter, Jo Nova and retired meteorologist, William Kininmonth.[85]

While a relatively small lobby group, with a budget of just over $2 million, the IPA influences public thinking through a constant stream of free opinion columns to copy-hungry media outlets from the *Australian Financial Review*, *The Australian*, *The Age*, *Herald Sun* and the ABC's 'The Drum'. Just as influential though are the regular events, whether hosting sceptics or launching reports, where the think tank's 20 staff mix with Liberal politicians, conservative media commentators and media owners.

The IPA's power network was on display at its 70th anniversary dinner in Melbourne in April 2013. The master of ceremonies for the event was Andrew Bolt, with keynote address by Rupert Murdoch, the owner of the majority of Australian newspapers, and another by Tony Abbott, the leader of the Opposition. Others attending were the head of News Corporation in Australia, Kim Williams; Executive General Manager of Ten Network, Russell Howcroft; Victorian Premier Denis Napthine; former mining executive, Hugh Morgan; Gina Rinehart and Cardinal George Pell.[86] In his speech, Abbott responded to an article in the IPA's magazine urging him to be bolder by adopting the think tank's 75-point plan for government. Abbott was keen to please:

> I want to assure you that the Coalition will indeed repeal the carbon tax, abolish the Department of Climate Change, abolish the Clean Energy Fund ... We will abolish new health and environmental bureaucracies. We will deliver $1 billion in red tape savings every year. We will develop northern Australia. We will repeal the mining tax. We will create a one stop shop for environmental approvals ... So, ... that is a big 'yes' to many of the 75 specific policies you urged upon me in that particular issue of the magazine.[87]

What think tanks such as the IPA and like-minded groups effectively do is to free up the coal industry's primary lobby groups, such as the ACA and the companies themselves, from damaging their own credibility by being publicly identified as rejecting the scientific consensus on climate change. In this way the climate change sceptics can publicly pursue Plan A of ridiculing any action on climate change with whatever the fashionable argument of the day happens to be, while the coal companies and their lobby groups can pursue Plan B of negotiating with the government to design a response that has minimal effect on them.

Central to the plan, however, is keeping obscure any financial connection between the power generators, coal miners and the think tanks and climate change sceptics. Back in May 2001 IPA's Director of Deregulation, Alan Moran, appeared before a Productivity Commission inquiry into the electricity industry and explained that 'about a dozen energy firms' contributed sufficient funding to run the IPA's Energy Forum. The forum, Moran stated, was 'a round-table discussion shop on one part, and on another part as a sort of advocacy unit'. Moran did not specifically state the names of the dozen companies, but volunteered that 'there are occasions when we may take positions which are somewhat different from those of the funders', but conceded 'obviously that doesn't happen too often, otherwise they'd stop funding us ...'[88]

Appearing at a Senate Inquiry in 2009, Moran confirmed that funding for the Energy Forum came from firms in the electricity and gas industries, but avoided naming which ones.[89] However, there are some clues. For the best part of eight years until late 2010, the chief executive of Loy Yang Power, Ian Nethercote, was a director of the IPA. For its part, the IPA was a loyal supporter of all things Loy Yang, a company that is now a wholly-owned subsidiary of AGL. When the carbon tax was proposed, Moran bemoaned in a *Herald Sun* opinion column that it would destroy the asset value of companies like Loy Yang, or possibly already had.[90]

The IPA has also carved out an additional role; peddling doubt about the viability of alternatives to coal-fired power stations, and criticism of government programs aimed at promoting renewable energy or energy efficiency measures.[91] In particular, the IPA has played a behind-the-scenes role of fostering opposition to wind farms, which threaten the market share of incumbent coal power generators. Early on, two IPA activists were quietly working away on a committee advising one of the key anti-wind farm groups, the Landscape Guardians.[92] On another occasion Moran was a speaker at an anti-wind farm conference of the Australian Environment Foundation (AEF),[93] an IPA spin-off group. When an inquiry was held into proposed noise and other regulatory restrictions on wind farms, the AEF and local affiliates of the anti-wind Landscape Guardians were out in force.

One person drawn into opposing wind farms is Maurice Newman, who has a 420-hectare farm at Crookwell in the Southern Tablelands of NSW near where a wind farm is currently being expanded. One of the founders of the Sydney-based right-wing think tank, the Centre for Independent Studies, Newman bought his farm after a successful life as a corporate high-flyer. Newman loathes wind farms with a passion[94] and strongly supports a legislative move by Senator John Madigan of the Democratic Labor Party and Independent Senator Nick

Xenophon to curtail their expansion in Australia.[95] If their bill were to win approval, one of the rapidly emerging threats to domestic coal-fired power would be largely nobbled.

Newman's anti-wind farm activism is in many ways indicative of climate change sceptics who have morphed into anti-wind farm activism. When he was Chair of the Australian Broadcasting Corporation, Newman infamously made a speech to ABC staff claiming that climate change sceptics had been censored, despite 'a growing number of distinguished scientists ... challenging the conventional wisdom with alternative theories and peer reviewed research'.[96] Newman's claims about climate science, which recycle the arguments of the climate change sceptics, would be of no great consequence in his post-ABC life were it not for Tony Abbott.[97] In late 2012, Abbott – who a few years earlier famously described climate change as 'absolute crap'[98] – announced that if he becomes prime minister he would appoint Newman as chairman of a new body, the Prime Minister's Business Advisory Council. Along with unspecified representatives from the resources sector, the council will be tasked with meeting Abbott regularly to 'address issues of concern to the business sector'.[99] Announcing the appointment, Abbott pledged that he would 'end the bad blood between government and business' and 'talk to people before we make decisions'. Who better than a climate change sceptic, neo-liberal think tank founder and anti-wind property owner to help the government implement its deregulation agenda?

||||

It is precisely such close behind-the-scenes cooperative relationships between government and industry that the coal and power generation industries have enjoyed for so long. Having built up substantial political capital with the Liberal–National Party by helping cripple the Labor government over its plans for an emissions trading scheme, a super profits tax and the carbon tax, Big Coal is set to reap even bigger rewards if Abbott wins the federal election and implements his plans to scrap the carbon tax, have a 'serious review' of the Renewable Energy Target, slash 'green tape' that could hamper new coal projects and revert to foot-dragging in international climate talks.

6 'CLEAN COAL' RUSE

> 'I don't believe in clean coal ...
> I think we should stop wasting time on it and money.'
> BOB CARR – Minister for Foreign Affairs and
> former NSW Premier [1]

For longer than most of us probably remember, Australian politicians have talked up 'clean coal', beating the drum routinely ever since climate change first peeked over the political horizon. Back in 1990, Bob Hawke confidently flagged clean coal technology as 'a possible basis' for coal exports into Asia.[2] His Energy Minister, John Kerin, assured Parliament that when it came to clean coal, 'as the world's foremost coal exporter, we will certainly be leading the charge'.[3] In 1994, with growth in Australian coal exports to developing country markets in mind, Industry Minister Peter Cook announced federal funding for 'clean coal techniques which will significantly reduce greenhouse gas emissions'.[4] 'Black coal utilisation: cleaner use, bigger export dollars', declared one media release, 'Black coal goes green', purred another.[5]

Yet, capturing carbon dioxide and storing it underground was the last thing on the minds of these clean coal advocates. Talk of carbon capture and storage (CCS) didn't start until the new century. For many years clean coal simply meant the superior quality of Australian coals, and the potential to use it more efficiently. It was a flimsy basis upon which to brand Aussie coal as clean coal, especially with the strong link being made even then between rising coal use and global warming. But it provided sufficient moral cover to successive Australian governments to allow them to justify a massive expansion in coal exports. In 1994, seemingly oblivious to the environmental consequences, Resources Minister David Beddall told a coal industry audience:

> We need to ask where we might be in 10 or 20 years time, and work backwards to the decisions we need to make now to achieve our goals … the Australian coal industry has two choices: 1) to maintain current practices, returns and levels of output; or … grow![6]

Twenty years later, the annual emissions generated by Australian coal are nearly twice what they were when Beddall urged the grow option.[7] No one seems to have thought to work back from that dubious achievement, or anticipated where the country that banked on clean coal might be now.

When Australia's government changed political colour from red to blue in 1996, the clean coal pitch continued. In his 'Safeguarding the Future' speech of 1997, leading up to the global climate treaty negotiations in Kyoto, John Howard stood proudly by the coal industry's environmental credentials.[8] 'Thanks to best practice efficiency standards in fossil fuel driven plants,' he said, 'Australian energy suppliers will be able to stand tall when it comes to being clean, green and cost competi-tive'. There was still no explicit mention of capturing and storing car-bon dioxide (CO_2) underground, nor was it mentioned in the Productivity Commission Inquiry on how to keep the coal industry competitive (with regard to the environmental objectives of governments). [9]

However, when former coal executive turned Resources and Energy Minister, Warwick Parer, gave a speech to an ACA conference in 1998 there was a hint that something better than coal quality and efficiency gains might soon be needed.[10] Parer spoke of 'new advanced clean coal technologies', and warned that unless they were deployed soon, it 'could effectively exclude coal as a viable energy source post 2010'. The incremental improvements and quality edge that had previously been packaged as clean coal might not cut it soon. Big Coal urgently needed a silver bullet – a big technological breakthrough. Or, as it turned out, it could adopt and adapt one that a fellow fossil fuel industry had been using for years.

Something old, borrowed and new

The idea of injecting waste CO_2 into otherwise inaccessible reserves to force oil and gas to the surface was something the petroleum industry had been doing since the 1970s. But its application to coal use seemingly took years to register. As late as the mid-1990s, CCS was not a noticeable part of the global coal industry's clean coal agenda. When, for example, the Coal Industry Advisory Board (CIAB) at the International Energy Agency (IEA) released a series of papers in 1996 canvassing the options being pursued by the coal industry, CCS wasn't even mentioned.[11] Yet climate change was now an existential threat to the coal industry worldwide, and without a credible explanation about how coal use might be decarbonised the industry's future was in doubt. The tantalising possibility that CCS could be the coal emissions panacea took hold as the industry realised that 'enhanced oil recovery' technology could be adapted to capture CO_2 from coal burning, convert it into liquid form and pump it underground. If CCS could provide a clean coal narrative deemed sufficiently credible, widely enough, the industry could keep expanding on the basis that the emissions could eventually be overcome. For coal industry spin doctors it seemed like an easy sell. With a wave of a magic high-tech wand, dirty coal could be re-birthed as clean coal. That it involved an existing technology made the idea more plausible.

Calls for government to fund and foster CCS grew progressively louder, with a great many people genuinely wanting to believe in its potential. The economic and political consequences of CCS not delivering were presumed to be catastrophic, and worth avoiding if at all possible. Yet, somehow CCS always had a 'too good to be true' ring to it. After all, for the technology to reduce total emissions from coal use worldwide, all new coal-fired power stations and steel mills would need to be built differently, or they would need to be retrofitted with CCS relatively soon, as would thousands of existing ones. Billions of tonnes of CO_2 would need to be piped long distances, turned into liquid form and then pumped to hundreds of communities willing to host CO_2 dumping sites, or alternatively it would have to be buried underneath the seabed. A vast network of pipelines would need to be installed to cater for the existing coal-fired infrastructure of Europe and North America and countries like Australia. An even larger network would have to be built in developing countries even faster to match the pace of growth in coal-based power and steel production there.

For all this to work on a global scale, even Exxon Mobil acknowledged that widespread deployment of CCS involves 'potentially duplicating the oil and gas industry's infrastructure – which has been built over 100 years – in a third of the time'.[12] Making the task even harder, the CCS process requires 25 per cent more energy than conventional coal use, so lots more coal would be used. Furthermore, if CCS was to be deployed on a commercial scale it had to be competitive without subsidy, and that required a CO_2 price of $100 a tonne, or perhaps more.[13] Once stored, the CO_2 would then have to remain stored for hundreds if not thousands of years. To be effective on a global scale, it had to be installed in countries like China and India, where large, new coal-fired facilities are being added every couple of days, and where there is no prospect of absolute emission reductions and $100 carbon prices.

From the very beginning of the global negotiations to combat climate change these countries made it clear that they had no intention of compromising their own economic expansion or of acting before the western countries responsible for the problem had led by example. While there were uncertainties, this was all staring us in the face when CCS was first widely put forward as a serious clean coal solution 15 or so years ago. It was an expensive long shot with an extreme degree of technical and political difficulty. Time wasn't on its side, neither was geography or economics, not to mention inertia. Thus, to believe in it took enormous gall, self-interest or hope-laden faith from people who genuinely wanted to believe all the huge obstacles could be quickly overcome.

Betting the future on CCS

Nowhere have the gall, self-interest and faith been in greater abundance than in Australia, where politicians led the charge with bipartisan recognition that Australia's future as a coal exporter depended on the CCS story being believed. The action heated up from 2001 after John Howard met George W Bush in Washington on the eve of the 9/11 terrorist attacks. Bush explained that the United States was working on an alternative to the targets and timetables approach of the Kyoto Protocol resoundingly rejected by the US Senate; a 'technology based protocol' emphasising clean coal technologies. Having promised to regulate CO_2 emissions during the 2000 presidential elections, Bush was now back-pedalling, with much of the pushing being done by a hand-picked darling of fossil fuel industries, Vice-President Dick Cheney. (Cheney was CEO and Chairman of oil and gas services giant Halliburton prior to the election, before being made Chairman of the new administration's Energy Task Force soon afterwards.) Howard returned from the meeting emboldened, determined that Australia would continue to be an 'energy superpower'. In the year that followed he chose World Environment Day to declare that Australia would not ratify the Kyoto Protocol, but instead pursue a technology-based alternative, working with the world's biggest emitters – many of which did not face emission reduction targets under Kyoto.[14]

Clean coal quickly became a central part of the narrative. Just as Cheney had invited fossil fuel industry heavyweights into his Energy Task Force to fashion a new energy policy that suited oil and coal interests, Howard's government invited their Australian counterparts into the Lower Emissions Technical Advisory Group (LETAG). Its input culminated in the *Energy White Paper 2004*, which made it clear that CCS was to be a central plank of Australia's climate change response.[15] Since then, federal and state legislation has been pushed through parliaments and billions of dollars of taxpayer funds earmarked for CCS research and demonstration. Close to $200 million was dedicated for CCS via the Low Emissions Technology Demonstration Fund, including $100 million for HRL's proposed 600 megawatt power station in the Latrobe Valley, and another $50 million for a flue gas capture plan at Callide in Central Queensland. Tens of millions of dollars were directed to the CO_2 Cooperative Research Centre (CO_2CRC) in Victoria. Australia also took a leading role in pulling together international support for CCS, through the Asia–Pacific Partnership on Clean Development and Climate (Bush's technology-based alternative to Kyoto), and through a new Carbon Sequestration Leadership Forum (CSLF).

Having not even mentioned the technology in his 'Safeguarding the Future' speech prior to Kyoto, just a few years later Howard was adamant about its viability, saying, 'Carbon capture and storage projects such as this are crucial to the future of Australia's important coal industry, and to the national and global challenge of reducing our greenhouse gas emissions. Australia must aspire to be a world leader in clean coal technology'.[16] Along with nuclear power, Howard made it clear, 'The answer is a greater emphasis on clean coal'.[17]

His enthusiasm was shared by Industry Minister Ian Macfarlane and Environment Minister Malcolm Turnbull. In a 2007 joint statement they said, 'adopting clean coal technology will be one of the most important actions Australia takes to address greenhouse emissions'.[18] According to Turnbull, 'Australia's progress towards cleaner coal is one of the most important ways we can make a global contribution. The coal industry is vital for Australia and vital for the world.' Foreign Minister Alexander Downer warned, 'make no mistake, if we do not find good solutions which allow for the ongoing use of coal while avoiding the release of CO_2 into the atmosphere, particularly in China, we do not have a solution to managing global emissions and avoiding the worst impacts of climate change'. In effect, if CCS didn't deliver, expanded coal use would lock in worst-case scenarios.

Labor's enthusiasm was equally strong. Successive federal leaders from Beazley to Latham to Rudd made it clear that CCS would be just as high a priority for any Labor government. Beazley called clean coal technology one of the two transformations crucial to an effective climate change response by Australia in a blueprint released in 2006. 'We have 300 years' worth of coal reserves in this country', he said afterwards. 'There is always going to be a requirement for it so ... we have to make sure that industry can progress within an environmentally friendly context'.[19] Labor Resources spokesman, Martin Ferguson, was of a similar view: 'obviously with a heavy reliance on coal, especially the difficulties of brown coal in Victoria, you have to invest in clean coal technology'.[20]

At state level, with the reticent exception of then NSW Premier Bob Carr, Labor was similarly keen. With carbon storage in the Bass

Strait in mind, Victorian Premier John Brumby said in 2007, 'Our government is investing heavily in clean coal technology and the Latrobe Valley is well placed to benefit from developments in all aspects of the clean coal cycle'.[21]

Nowhere was Labor's enthusiasm more determined than in Queensland where Peter Beattie and Anna Bligh found over $300 million dollars for clean coal. Setting up the Centre for Low Emissions Technology (CLET) in 2003 with $9 million from his government, Beattie's vision for CCS was clear early on: his 'smart state' would lead the world in clean coal technology demonstration and deployment, and in doing so, develop expertise that its universities could sell around the world – locking in the future of the state's growing coal export industry in the process. Queensland wouldn't just export coal to China and India; Beattie planned 'a new export industry, with more jobs, exporting our clean-coal technology'.[22] Of CCS, he said in 2007, 'This is the only way to go. People now accept the reality of clean coal technology'.[23] In an interview on '60 Minutes' he said, 'What we have got to do is clean it up. And the way to do that is clean-coal technology, zero emissions ... The technology is five years away, although hopefully we can do it sooner. But that will change the planet.'[24]

Fellow Queenslander, Kevin Rudd, was equally keen. Without CCS Rudd could hardly expect to maintain the climate change action man image carefully fashioned in Opposition, while also backing rapid expansion of the coal industry. He came to office promising a new $1.5 billion clean coal fund, saying:

> We have an overwhelming national imperative at stake here – how do we place long term coal exports on a secure long term footing and the challenges which are emerging in terms of the climate change debate globally, and in terms of carbon emissions going forward? ... The way we do it is to bring about a new clean coal initiative[25].

Ratifying Kyoto helped, but CCS was the climate-friendly alibi Rudd relied on most as he strutted the global stage masquerading as climate change crusader. Pivotal to Rudd's CCS agenda would be the position of Ross Garnaut, an economist and former mining executive appointed by Rudd as the Australian government's climate change adviser. After many months of research, consultation and deliberating, Garnaut delivered the answers. While acknowledging the uncertain economics of CCS in his 2008 report, Garnaut saw no reason not to project large increases in demand for Australian coal for many decades to come. He was sufficiently confident (or perhaps hopeful) about the prospects of CCS to suggest that its commercial deployment could see Big Coal's power generation heartlands become 'regions of strong expansion and prosperity' and that Australian government should 'make substantial commitments' to support industry's CCS research plans.[26] The contingency that CCS might not work got less emphasis.

Like so many others, Garnaut seems to have genuinely hoped the technology would deliver on all its hype soon. He also understood what was at stake, saying: 'the future of the Australian coal industry depends critically on the success of carbon capture and storage not only in Australia, but in the rest of the world, and especially in Australia's major coal markets in Asia'.[27] However, he raised no ethical objection to selling ever-increasing quantities of coal while the world waited to find out if CCS would work on scale and in time. Odds were it would take decades to find out, so deliberately expanding coal production while not knowing whether CCS would deliver or not was a massive gamble with huge environmental risk. But it was a gamble with which Garnaut, Rudd and so many others seemed comfortable. On the basis that CCS would work, Rudd bankrolled new coal export infrastructure. There was $580 million for duplicating railways in order to double coal exports from the Hunter Valley.[28] In parallel, money also poured into proving up CCS; $1.9 billion was earmarked for the Carbon Capture and Storage Flagships Program to fund between two and four 1000 megawatt CCS-equipped coal power stations. A further $370 million was earmarked for the National Low Emissions Coal Initiative, which included, among other things, mapping potential carbon dumps, and funding a new agency called Australian National Low Emissions Coal Research and Development, which would seek cost reductions for 'low emissions' coal, and funding for research on geological storage of CO_2. The federal government also established the National CCS Council with representatives from the coal, gas and petroleum sectors, whose role included dispelling community concerns that might prevent public acceptance of CCS.[29]

By late September 2008, the CCS bandwagon had gathered significant domestic momentum. 'We have got to crack the whip and make it happen,' Rudd had said, announcing the establishment of the new the Global CCS Institute, based in Canberra.[30] It would receive $100 million a year, he said, because 'we have decided that we need to have some skin in the game'.[31] 'Climate change is a threat for the future. It is a threat also for the future of our coal industry in Australia,' he said.[32] In response to a sceptical question, Rudd set out what mattered to him: 'We're talking about Australia's largest export here, which generates tens of billions of dollars worth of revenue each year ... I'm on about long term protection of the revenue base'.

At the G20 meeting in L'Aquila Italy in mid-2009, Rudd effectively used the global stage to re-announce the Global Carbon Capture and Storage Institute (GCCSI). Saying the time for talk about CCS was over, he said he had set up the GCCSI 'to get large-scale carbon capture and storage projects done around the world, not just talked about'.[33] British businessman Nick Otter was recruited as the Institute's CEO, and an International Advisory Panel was set up in order, said Rudd, to 'raise the international profile and credibility'. It included former head of the World Bank, James Wolfensohn as Chair, with former UK Treasury climate change adviser Sir Nicholas Stern as a member.[34] The future was looking bright.

All aboard the bandwagon

As CCS rode a wave of federal government backing, and with the strong support of the Queensland, Victorian and WA state governments, a wide range of organisations and prominent individuals beyond politics keenly climbed aboard the bandwagon. There was a genuine desire on the part of many to believe that CCS might offer Australia a way to have its coal and export it. Not surprisingly, coal companies fed the sense of optimism. Millions of dollars were channelled into Co-operative Research Centres, CSIRO and other institutions keen to capitalise on CCS research and development opportunities. In 2006, the coal mining industry had announced a formal arrangement to support CCS via its Coal21 Fund. Under the plan, coal companies would contribute 20 cents a tonne of exported coal, an estimated $1 billion over 10 years. The stated aim of the fund was 'to demonstrate the technical and economic viability of low-emissions coal technology, leading to commercial deployment by 2015.'[35] Energy companies, many of them state-owned electricity utilities, talked up the potential of CCS and joined consortia looking to demonstrate the technology. The ESAA issued a report claiming that clean coal (and nuclear) was the way to go if Australia was going to cut emissions at least cost. In nationwide advertising, and on a flashy new website laden with scientific experts from CSIRO and elsewhere, the industry very effectively communicated the notion that CCS had to work, that it would work, and soon.[36] It appeared that industry was putting big money behind its rhetoric to demonstrate CCS technology as a precursor to commercial deployment on a large scale. The sum of $1 billion dollars from the coal industry sounded impressive enough, although in reality it represented only a tiny percentage of the industry's annual profits, and as coal prices rose, the per tonne contribution to CCS stayed the same.

Coal unions got behind the push, with the country's largest coal union arguing that, 'The CFMEU is not interested in palliative care options for the industries our members work in'. Instead, it foresaw a bright future for the industry if efforts were dedicated to 'transitioning the coal and power industries to become low carbon industries through CCS'.[37] As CFMEU National President Tony Maher put it,

> Rapid demonstration of CCS in Australia is essential to securing employment prospects in regional Australia – jobs in coal mining and jobs in new high-tech CCS power plants. Mineworkers know that their industry and their jobs only have a future if coal use – and gas use – becomes a low emission industry here and overseas. And with coal being Australia's largest export industry, we need to lead the way in the development of that technology.[38]

Joining the CFMEU and the ACA in 2008, some environmental groups lent credibility to CCS by forming a coalition to demand yet more federal support and funding for it. As well as demanding a new National Carbon Capture and Storage Taskforce, they wanted the federal government to subsidise at least three 500 megawatt demonstration projects, a favourable legislative regime, an 'education campaign', and the mapping of suitable CO_2 dumping sites and a 'pipeline investment framework'.[39] Announcing the partnership, WWF-Australia CEO Greg Bourne said, CCS was 'needed if we are to avoid dangerous climate change'. Hinting that Peter Beattie was right in saying CCS might become a new export industry, Climate Institute boss John Connor said, 'Australia's leadership in the development of CCS can also contribute to emissions reduction in emerging economies such as China and India'. It was a line eagerly picked up by sections of the media, including *The Australian*, which began editorialising that with nuclear and a carbon trade 'acceptable' to Australia both unlikely, 'exploring clean coal and geosequestration technologies makes far more sense for a country with our resources profile'.[40]

A stream of eminent scientists and research institutions further raised hopes along the way. Commonwealth Chief Scientist Robin Batterham reassured federal and state energy ministers that the entire process of CCS would cost around $10 a tonne.[41] That the part-time Batterham's other job was as a chief technologist with Rio Tinto, one of Australia's largest coal miners, raised eyebrows, but caused Howard little concern. Meanwhile, ABARES, at the time the federal government's principal source of economic advice on climate policy, fed the idea that CCS could deliver large-scale abatement as early as the 2020s.[42] Including CCS among technologies that 'either currently exist or are realistically achievable, they modelled various scenarios from 2010 assuming the availability of CCS worldwide. Once CCS was economically feasible they foresaw 'around two-thirds of the (coal- and gas-fired electricity) industry adopting the technology after ten years'.[43] The public statements of various research bodies sang an even more positive chorus. CSIRO's Energy Futures Forum, for example, said CCS is 'not only crucial from the coal industry's perspective, but is also potentially one of the lowest cost abatement options for Australia'.[44] Even Australian of the Year, Tim Flannery, having previously warned that coal might become the asbestos of the future, now said there was 'little choice but to pursue a solution that involves (clean coal)'.[45] There was no alternative, he suggested, given the pace at which coal-fired power was being added in countries like China, and because 'China will not simply knock down its newly constructed power plants...'.

While it went largely unnoticed, perhaps because many of the hopes were genuine, the CCS-boosting institutions were often heavily funded by companies with an interest in the CCS story being bought. In various instances, including a few at CSIRO, the same vested interests were also generously represented on advisory boards overseeing and guiding research work. Meanwhile, senior government officials, including departmental heads, took to chairing boards of the organisations that most avidly promoted CCS.⁴⁶ Tim Flannery was by this point an official adviser to Siemens, a strong CCS backer that is selling ever-more conventional steam turbines for use in coal-fired power stations (especially in developing countries), and Tata Power, India's largest private power company, which has big plans to expand coal-fired production.⁴⁷ Not surprisingly, in this sort of echo-chamber, optimism flourished about CCS delivering at home, and becoming a new export industry. Australia became a honey pot for the world's CCS experts, a bonanza for anyone with expertise and enthusiasm for the technology.

International traction

Internationally, the drumbeat was similar, albeit heavily concentrated in just a few countries.

The conversion of the Intergovernmental Panel on Climate Change (IPCC) was typical. In its first report in 1990, which rang the alarm bells on greenhouse gas concentrations, it was vague and cautious about proposals to bury carbon underground and/or the cruder and cheaper option of just pumping it into the deep ocean and hoping it would stay there.⁴⁸ Five years later, the IPCC added a little more detail about the possible technologies, but added heavy caveats about the significant increased costs of 50–80 per cent for gas and coal power and cautioned that using CO_2 to boost oil and gas production might only reduce human greenhouse gas emissions by about 1 per cent at best.⁴⁹ As the global network of CCS boosters grew, however, the IPCC gradually warmed to CCS as being one possible, albeit expensive, solution. In 2005, undoubtedly mindful of the CCS-spruiking of countries like Australia, Canada and the United States, the IPCC agreed to issue a *Special Working Paper on Carbon Capture and Storage*. Ostensibly the report was an impartial synthesis of the science according to the world's leading experts in the field. In practice, it was part glowing justification for government faith in the technology and part job application for researchers accustomed to obscurity and now suddenly centre-stage. Perhaps not surprisingly, given that it was authored largely by avid CCS proponents, some of whose positions were partly funded by coal or coal-fired power companies, the *Special Report on CCS* was a far cry from the more cautious posture taken by the IPCC in earlier reports. It was a coup for CCS boosters and a turning point for the IPCC, which from then on routinely included CCS as part of its narrative on how to respond to climate change.

Heavily influenced by the *Special Report on CCS*, Sir Nicholas Stern in his 2006 assessment of climate change proclaimed that 'extensive' use of CCS would allow 'continued use of fossil fuels without damage to the atmosphere.'[50] CCS, he declared, was 'essential to maintain the role of coal in providing secure and reliable energy for many economies'. It could provide 20 per cent of the emission savings required by 2050, he said, and without it the cost could be 60 per cent higher.[51] Even though Stern somewhat moderated his belief in CCS a few years later, others embraced his initial enthusiasm. The US-based Carbon Mitigation Initiative, for example, a project of Princeton University, part-funded by Ford and long-time CCS proponent BP, popularised what became known as the 'wedges' approach to reducing carbon emissions. The Princeton project identified CCS as one of 15 key strategies that could deliver savings of a billion tonnes of carbon annually. On a graph, each of these could deliver a wedge, and seven stacked wedges could provide the change required to stabilise global emissions by 2050 (this was later revised upwards to nine). On the heroic assumption that CCS could be equipped to 800 coal-fired power stations by 2050, as well as generating emissions-free hydrogen and transport fuels in vast quantities (delivering up to three of the required wedges in all), the CMI work made CCS look essential. For many, this became an article of faith for years afterwards.

For its part, the IEA was also rapidly assuming the role of global CCS cheerleader, from 2006 onwards churning out over 20 reports, briefing papers and background papers that extolled the virtues of CCS and the need for public funding. CCS, it claimed, was one of a number of 'zero emissions technologies' which 'will virtually eliminate emissions from the conversion of fossil fuels and their consequential pressures on health and the environment'.[52] While the IEA trades on the perception that it is a think tank dispensing sage advice to its 27 member governments, the reality is that it is close to the coal industry. For the last 30 years it has organised the CIAB, comprising 'high level executives from coal-related industrial enterprises' to 'provide advice to the IEA on a wide range of issues relating to coal'. Australian coal lobbyists from Xstrata, Rio Tinto, BHP Billiton, Delta Electricity and the ACA are among the 45 executives invited into the IEA's inner sanctum. The IEA's vision of a coal-friendly future makes the Princeton/BP projections look decidedly timid, proposing as realistic 100 CCS projects implemented by 2020 and over 3000 by 2050. Under this scenario, CCS could provide one-fifth of the emission cuts required to halve global emissions relative to 2005 and keep atmospheric CO_2 concentrations at 450 parts per million.[53]

CCS projects galore

As political and institutional support for CCS expanded, grand plans for CCS projects also proliferated. Dozens of projects were announced. By 2009, the Global CCS Institute was pointing to a pipeline of 247 CCS projects around the world, 34 of them completed.[54] There were 213 active or planned projects listed, including a long list of integrated projects at coal-fired power facilities that would capture about 100 million tonnes of CO_2 per annum once completed, and most of which would be up and running by 2015.[55] Granted, 100 million tonnes of CO_2 is less than half of 1 per cent of the annual emissions generated globally from coal use now, nearly 90 per cent of the projects cited were in developed countries (mostly Australia, the United States and Canada), and only 10 per cent were in China and India, where the vast majority of new coal-fired power capacity and steel mills are being added.[56] Even so, it still sounded serious, just as it did in Australia where CCS projects were being announced with great fanfare on a regular basis.

In Queensland, the state-backed $4.3 billion ZeroGen project proposed a state-of-the-art 400 megawatt coal-fired power station west of Rockhampton, capable of capturing and storing 60 million tonnes of CO_2 over the life of the project. According to ZeroGen's proponents, it could accelerate 'rapid deployment of low emission coal technology at a cost that will preserve Queensland's competitive advantage as a power generator and ensure the continued mining, export and use of Australian black coal'.[57] Peter Beattie called it 'one of the most important CCS projects, not just in Australia but in the world.'[58] In WA, BP and Rio Tinto announced a $2 billion hydrogen energy coal-powered plant for Kwinana.[59] BP said it would be 'an industrial-scale' CCS project that 'would generate enough electricity to meet 15 per cent of the demand of south west Western Australia, while each year capturing and permanently storing about four million tonnes of carbon dioxide' (underground off the coast between Fremantle and Rottnest Island). It had 'enormous potential to affect the way that coal will be used for power generation across the world'. WA Energy Minister Francis Logan was clear about the implications, saying: 'the global deployment of clean coal technologies such as this has the potential to secure a long-term future for the export of coal'.[60]

In Victoria, Anglo American and Shell announced the Monash Energy Project, a plan to turn brown coal into a synthetic diesel and, once the project reached commercial production levels, to divert the CO_2 generated in the manufacturing process offshore for underground storage – most likely in the Otway Basin. 'The future starts now' said a glossy brochure from the proponent, and 'Clean coal is not just a pipe-dream'.[61] An upbeat Victorian government projected CCS demand equivalent to '15 million tonnes per year from 2015 on, mostly from the Monash project'.[62] The Energy Minister, Peter Batchelor, said that such projects had the potential to turn Victoria's 'unfathomable bounty of brown coal' from a 'heavy burden' to 'our supply of clean energy for the near future'. Meanwhile, in South Australia, Santos, Origin and Beach Petroleum announced plans to create a massive CO_2 storage hub at Moomba. Billed as 'a project for our time', Moomba Carbon Storage would 'store up to 20 million tonnes of CO_2 per annum and 1 billion tonnes over the life of the project'.[63] The proponents argued that it would 'deliver Australia its first large CO_2 abatement option, providing the storage solution which will see CO_2 from sources in eastern Australia transported to central Australia and permanently stored underground'. The project was to 'establish Australia as a leader in climate change initiatives and will provide a crucial enabler for the development of clean coal technology in eastern Queensland and New South Wales' Hunter Valley'.[64]

Between the political support, and projects being announced from east to west, CCS looked serious. Flashy PowerPoint presentations showed maps with dozens of CCS projects happening or in prospect, most of them handily located near geologically suitable storage sites and existing gas pipeline infrastructure. Some slides showed where the pipelines would head offshore to other nearby storage under the seabed. Even longer pipelines were shown snaking off into Central Australia from South Australia, the Latrobe Valley, Wollongong, the Hunter Valley, and South East Queensland. Just as inland Australia's rivers drain benignly into Lake Eyre, the country's CO_2 would drain safely in the same direction to the Moomba underground storage facility. New developments at each of these projects triggered another round of sunny publicity – a new project partner, a government grant, new holes drilled, a sod turned on-site – all requiring press releases, politicians to cut ribbons, joint press conferences and such like. With so much activity, it seemed that CCS was just around the corner.

The sound of hot air freezing

Gradually, however, the gap between the promise of CCS and the reality began to test public faith in the technology and strain credulity. Timelines repeatedly slipped, project costs escalated, and very little CO_2 was actually being captured and stored. The core problem was the technical and financial viability of turning the enticing idea of CCS into a reality. While CCS boosters could say the technology had operated for years in the oil and gas sector, and that most elements of coal-based CCS were demonstrated, this was only true of small-scale pilot plants, usually involving one or two parts of the process – not the whole process. Integrated commercial scale demonstration was a much harder proposition. The first major hurdle was that CCS plants are prohibitively expensive to build and operate; around 75 per cent more expensive than their CCS-free counterparts.[65] CCS enthusiasts argued that the high cost for 'first movers' – the initial plants – would fall as expertise was gained, and that the cost could be partly avoided through 'Enhanced Oil Recovery'. If CO_2 could be sold as a valuable by-product to companies keen to re-pressurise a declining oil or gas field, power stations might avoid CO_2 disposal and storage costs. Failing this, CCS enthusiasts assumed carbon prices would be in place and rising rapidly, rendering the technology attractive relative to conventional coal-based plants as well as gas and renewables.

Reality is panning out differently: carbon prices haven't changed the equation anywhere near enough and seem highly unlikely to any time soon. By most estimates, CCS is not commercially viable until carbon prices reach $100 – a long way off. This is the main reason for federal Treasury estimating CCS will not begin to be deployed commercially in Australia until the mid-2030s, and later in the developing country markets where most new Australian coal exports will go. The recent announcement to link the Australian carbon price with the EU (where carbon prices are lower than forecast here) pushes the CCS viability timetable out even further into the mid-2030s and beyond. Ironically, the coal industry's strident lobbying campaign against the carbon tax slowed the pace at which CCS might become viable by encouraging the government to adopt measures that minimise the carbon price and the speed at which it rises. Under the circumstances, the assumption that there will be sufficient demand for CO_2 storage within the next two decades to warrant large-scale investment looks dead.

Without that demand it's unlikely anyone will invest in the vast pipeline and other infrastructure required to accommodate CCS on a large scale. Even if the demand does emerge, and it is matched with suitable sites, building the infrastructure to transport the CO_2 would take many years. According to the federal government's own estimates, a pipeline over 300 kilometres long could take from three to six years to build, factoring in environmental approval and other design and investment decision processes. The Gippsland Basin in Victoria, for example, is viewed as the 'highest technically ranked storage basin' with 'the lowest transport and storage cost', but even with these advantages the government does not expect the infrastructure required to start storing CO_2 to be in place before 2022 at the earliest.[66] And 300 kilometres would be a small pipeline relative to most required for large-scale CCS in Australia. According to the federal government's own Carbon Storage Task Force, the most suitable locations for CO_2 storage in Australia are anything but conveniently adjacent to where most emissions occur, with the relative exception of Victoria. In South East Queensland the best location (the Euromanga Basin in the state's west) is 'more than 1200 km from emissions hubs'.[67] The Surat and Galilee basins are posited as 'stepping stones', but even these are 400–600 kilometres away. In NSW, the situation is probably worse. The emissions are concentrated near Sydney, the Hunter Valley and Illawarra, where 'the majority of the basins have low storage capacity'. Hope is held for the Darling Basin, but it's over 750 kilometres away, and might only take 5 million tonnes of CO_2 per annum; nowhere near enough to make a noticeable dent in the state's emissions (158 million tonnes per year). Unless the capacity estimate increases, emissions might need to be sent 1000 kilometres southward to the Gippsland Basin in Victoria, or even 1700 kilometres north into Queensland.

With such distant storage sites and such long pipelines, some running close by uneasy residents, all sorts of liability issues hamper large-scale CCS. In high concentrations CO_2 is a potentially fatal, toxic gas; in 1986, for example, a volcanic eruption caused CO_2 trapped at the bottom of Lake Nyos, Cameroon, to leak to the surface, killing living things within 25 kilometres, including 1700 people and their livestock. With CCS pipelines potentially coming much closer than this to nervous Australian communities and covering such long distances, proponents want to offload responsibility for any such disasters. Just who owns the geological formation underground into which the CO_2 is injected, and who owns the CO_2? Who is responsible if the CO_2 leaks from the pipeline or from storage, and what if things go wrong after the companies involved have gone out of business? In Australia, as in the United States and Europe, CCS proponents have sought to offload the expensive liability risks onto the taxpayer in order to minimise their own costs. In WA, for example, Chevron succeeded in persuading the state and federal governments to indemnify the massive Gorgon LNG gas project from any legal liability for CO_2 escapes that occur after the injection point is closed.[68] While no estimate of the potential liability has been released, hundreds of millions of dollars would be a conservative estimate for the projected 144 million tonnes of CO_2 to be buried.

Another problem is the amount of water required to accommodate large-scale CCS. A consultancy report undertaken for the National Water Commission in Australia has cautioned that CCS would place additional demands on already limited water availability.[69] Already, coal-fired power stations require vast amounts of water for cooling. The report noted that 'coal-fired power plants incorporating carbon capture and storage (CCS) could be one-quarter to one-third more water intensive' than existing water-cooled plants. By some estimates CCS could be even thirstier, with the cooling water requirement potentially doubled once CCS is added.

Then there are other hurdles. In developed countries like Australia, demand for electricity is levelling off thanks to a range of factors, from the proliferation of rooftop solar and insulation installation to the offshoring of energy intensive industry, to people cutting back on electricity use because of rising prices. As demand stagnates and wholesale electricity markets flatten, coal-fired power generators are becoming less keen on massive new capital outlays. They have little appetite for investing increasingly scarce shareholder funds into new conventional coal-fired plants, let alone CCS projects likely to be uneconomic for decades. They also know that if CCS makes no sense for new power stations, the idea of retrofitting it to ageing existing plant with a limited life is even more ludicrous. Under the circumstances, banks and other private investors are averse to providing significant funds. Meanwhile, coal companies are relatively more cash-strapped than they were a few years ago, thanks to declining prices and dampened investor enthusiasm for new coal projects. Between that and the fact that most of their coal is for export into markets where there is little pressure to reduce emissions, companies are less and less inclined to invest billions of dollars to build CCS plants that will do little more than demonstrate CCS is uneconomic.

With CCS looking increasingly unviable from all aspects, its shrinking supporter base has been left with few options but to rely on government funding. In developed countries they argue that substantial taxpayer funding is needed to carry all or a substantial part of the costs of the 'first movers' and much of the pipeline infrastructure and mapping of underground carbon dumps. As costs decrease over time, the argument goes, fewer government funds would be needed, especially as carbon prices rise, and if CCS plants can be treated as renewable energy-equivalent. The rhetoric is uncannily like calls for subsidies from renewable energy producers looking for a leg-up – calls the coal industry has steadfastly opposed and ridiculed. The CCS boosters, however, argue that coal is a special case – that it is strategic, of national importance because of the economic significance of coal-fired power and coal exports to the economy. Until CCS is an 'off-the-shelf technology', they argue, carbon prices won't work on their own. In addition to 'market pull' provided by carbon prices, they say, government support is required to provide 'market push' to get CCS onto the shelf.[70] For a time, these lines washed with government, and while optimism reigned, the public purse was easily pried open. But, as the global financial crisis eroded demand for new power stations and public finances alike, the headwinds got stronger.

Internationally, the prospects of CCS have also turned bleak. With little pressure to reduce emissions in developing countries, and conventional coal plants being rolled out there every few days, the technology is least viable where it is most urgently needed. CCS proponents would like subsidies from institutions like the World Bank and the Asian Development Bank, but with both content funding conventional new coal plants in the developing world that seems unlikely. CCS boosters have had a big win in having CCS plants deemed eligible to generate carbon credits that might be traded under the UN's Clean Development Mechanism (a 'flexibility mechanism' established to enable emitters in one country to purchase offsets for their emissions in developing countries).[71] But the win on carbon credits could turn out to be hollow, because in late 2012 the UN confirmed that conventional new coal-fired power stations in these countries without

CCS would be also eligible for CDM credits, as long as they were more efficient than what might otherwise have been built. While controversy rages over whether the emission savings being credited are even real, the rush to generate carbon credits from new coal-fired power without CCS (over 40 in China and India have sought CDM approval) further retards investment in the technology.[72]

The pennies drop

Things were supposed to be different: a higher carbon price, more demand for CO_2 storage much sooner, more coal industry investment in CCS, more global pressure to reduce emissions – not less. By now, CCS demonstration plants were meant to have shown beyond doubt that the technology was technically doable and could be up-scaled quickly. As none of this applied, the pennies have dropped for almost everyone who was once keen to back clean coal's Holy Grail. These days the maps dotted with CCS plans that were used just a few years ago to prove the case, do the exact opposite. One after another the CCS projects that featured most prominently have been completely cancelled or indefinitely shelved.

In 2008, the Monash project in Victoria went on ice and the Kwinana project in WA fell over as it became clear that the geological formations off the Fremantle coast assumed to be suitable for storing CO_2 were anything but. In 2009 the Moomba CO_2 storage hub in Central Australia was canned, with proponents blaming the lack of a carbon price required to make it viable. Suddenly the Hunter Valley and South East Queensland coal burning hubs, remote from any obvious CCS solution of their own, were without a CCS lifeline. Other projects went the same way. The proposed 450 megawatt Coolimba CCS power station in WA (meant to capture 2 million tonnes of CO_2 per annum) was canned in mid-2012, the Wandoan power project in Queensland (meant to capture 2.5 million tonnes per annum) is 'under review', and the HRL project in Victoria looks to have been indefinitely shelved. Most spectacularly, after burning nearly $150 million in taxpayer funds, the ZeroGen project was killed off in 2010 when the Bligh government cut its losses and abandoned it. The Gorgon CCS project in WA looks likely to proceed, but it doesn't involve coal-fired power.

Internationally, it's the same story. The 2008 decision by the US Bush administration to pull funding from the FutureGen project, only months earlier referred to as the centrepiece of the US clean coal effort, was like a starting gun for global CCS retreat.[73] Across the world, projects once promoted as the future became things of the past. The sobering reality, as IEA acknowledges as of mid-2013, is that 'No large-scale installations exist yet in electricity production'.[74] Meanwhile, on the horizon, the likely commercial-scale CCS projects are today few and far between. With a snails-paced roll-out, the promised big cost reductions from 'learning by doing' look illusory. A mid-2012 US Congressional Budget Office report found that without international agreement to cut emissions, subsidising CCS-produced electricity at home or abroad is likely to be futile. Largely because of its higher cost, they said, the technology might 'never prove to be cost-effective'.[75] The IEA's target of 3000 CCS plants by 2050 seems forlorn. Reality hit home in 2012 when BHP Billiton coal division Chief Executive Marcus Randolph told the *Australian Financial Review* that while investment in CCS was positive, 'he had his doubts that that would be enough to make coal a good choice for power generation in the longer term'.[76] Thus, he said,

> In a carbon constrained world where energy coal is the biggest contributor to a carbon problem, how do you think this is going to evolve over a 30- to 40-year time horizon? You'd have to look at that and say on balance, I suspect, the usage of thermal coal is going to decline. And frankly it should.[77]

Although BHP Billiton makes more money from metallurgical coal than thermal coal, coming from one of the main funders of the technology around the world, this spoke volumes.

As the supply of relatively cheap gas explodes and the cost of renewables falls, CCS increasingly looks like a technology whose chance to shine has come and gone. In the wake of spectacular undelivered hype, it has left even the truest believers despairing. As former CO_2 Cooperative Research Centre head, Peter Cook, frankly told the ABC in 2011 'There's no question we're at the crossroads when it comes to CCS. Time is not on our side.... we've just got to get large-scale projects underway and it's a concern that we're just not doing that'.[78] Another speaking out is Kelly Thambimuthu, the long-time Chair of the IEA Greenhouse Gas R&D Program, appointed by the Queensland government to oversee its clean coal effort in 2002. He quit in disgust in late 2010 after watching projects go nowhere, particularly ZeroGen. On the way out he took a swipe at the delays, saying, 'It's absolute nonsense for Queensland to export so much coal and make money from it and yet do nothing about reducing emissions.'[79] Interviewed later, he was even more blunt about the prospects, dismissing as unrealistic federal plans for commercial-scale CCS projects by 2020, 'There's nothing on the horizon in Australia that will take us to the 2015 and the 2020 target'. Asked whether he could put a timetable on large-scale CCS projects, he said,

> I can't because there aren't any serious power projects (involving CCS) out there that I could point to and say I can speculate what the timeframe is going to be ... If you built one of these plants from scratch ... it would probably take seven years to build a plant, and then you have all these uncertainties about pipelines and storage etc, so it's looking like a decade from now before you get under way.[80]

In late 2011, Chair of the Gillard government's National CCS Council, Dick Wells (a former head of the MCA) delivered Industry Minister Martin Ferguson a blunt assessment of the situation in Australia:

the carbon price is unlikely to be sufficient to pull through first-of-a-kind capture plants by the mid 2030s (as per Treasury's core scenario) without additional support from government ... (and) the carbon price is unlikely to provide the incentive to build oversized pipelines or commercial scale storage sites for some decades.[81]

Here were pro-CCS experts bluntly telling government that CCS won't happen by the mid-2030s without loads more public money. Tim Flannery, now chief of the federal government's Climate Commission, had also backed away from CCS. While foreseeing a 'healthy coal export industry' for 'some decades to come', he'd gone from being a 'great proponent' of CCS to believing it 'economically unfeasible'.[82]

Politically, bipartisan support for CCS has unravelled as reality has dawned. By 2009 as projects tumbled, Howard-era backer Ian Macfarlane retreated, 'You can talk about all the stuff you like about carbon capture and storage, that concept will not materialise for 20 years, and probably never'. The following day he added: 'the clean coal option has passed us by. Twenty years to wait before the technology is available. Thirty years before it is commercial. We will need to move on to other options by then.'[83] By the 2010 election Tony Abbott pledged that a Liberal government would axe funding for the GCCSI previously supported by his party. 'Frankly if we're the only country that is backing it and funding it, it's never going to happen, let's save the money'. More recently, and guardedly, when Abbott was challenged in an online forum by a CCS supporter as to why he didn't support the technology, Abbott was more deft, saying, 'I do support CCS...if it's viable but it's unlikely to be economic in Australia for many years'.[84] Even coal industry-friendly think tanks have begun playing down expectations with the IPA's Alan Moran, saying in mid-2011: 'in spite of vast outlays, there is no prospect that coal power based on carbon capture and storage will get off the drawing boards'.[85]

Labor still clings to CCS in its rhetoric, but the enthusiasm has waned. In 2011 Julia Gillard said confidently, 'The government is continuing to invest in carbon capture and storage. I believe it will be a big part of the way we deal with energy in the future and with carbon pollution in the future'.[86] Today, however, it is conspicuously absent from her speeches and media comments, and confidence has been replaced by sheepish hope. Dodging questions as to whether CCS was still a central plank of the government's climate response, and whether he still had faith in it, for example, Trade Minister Craig Emerson conceded to 'Meet the Press' in mid-2012 that the government wasn't factoring CCS into its emission targets for 2020. He added whimsically, 'If carbon capture and storage produces results that are commercially viable to further reduce emissions ... then that's a great thing'.[87]

The financial retreat from the technology in the post-Rudd era has been less ambiguous. In a series of federal budget announcements, the Gillard government has hacked away at CCS funding. Under cover of the floods disaster in 2011, for example, Gillard announced that $55 million would be cut from the GCCSI over the three years to 2014–15, with annual funding afterwards to be slashed to just $20 million a year. With the institute nominally earmarking $50 million a year for the direct funding of CCS projects around the world, Gillard also announced that $250 million would be cut from the CCS Flagships Program, though with $160 million of that allocated beyond 2015.[88] Meanwhile, a federal decision recently pulled funding from the HRL project in Victoria, when it failed to meet conditions placed on the grant, even after an extension. Under pressure from the Greens, and perhaps wary of throwing good money after bad the government has excluded CCS from funding under the new Clean Energy Finance Corporation, established as part of the carbon tax package. The retreat is mirrored at state level – most obviously the Bligh government's previously mentioned 2010 decision to abandon the ZeroGen project.

Some senior Labor figures still champion CCS, but the most zealous are increasingly realistic. In 2010 Peter Beattie noted: 'Queensland, Australia and the rest of the world have been working on clean-coal technologies for years. The problem is that where Australia is today is not a lot different to where we were in 1998'.[89] Former NSW premier, now federal Foreign Minister, Bob Carr, was more blunt, saying in mid-2010: 'I don't believe in clean coal. Where the power stations are in New South Wales, coal fired power stations, there aren't any empty aquifers. Clean coal is not an option or a solution there. I think we should stop wasting time on it and money.'[90] As one anonymous political adviser told The Age, within government, discussing 'clean coal has become an invitation for ridicule'.[91]

The mirage our taxes funded

Looking back, it's increasingly clear that the hype surrounding CCS has been a mirage largely maintained with carefully coordinated and mostly government-sponsored PR – much of it run out of Australia. From pilot CCS projects all over the world to expensive workshops and consultants' reports, to IEA programs promoting the technology, to setting up a network of sympathetic environmental groups internationally, things that have collectively looked like an international groundswell of support for CCS deployment in sync with the Australian position, turn out, on closer analysis, to have been subsidised by Australian taxpayers. So, for example, when Kevin Rudd visited a Beijing pilot plant in 2008 with an Australian media contingent in tow, it was sold as evidence of China's commitment to the technology. But what Rudd was showcasing in Beijing was made possible with $4 million of Australian government funding. This was an isolated example, but Labor's cashed-up GCCSI institutionalised the practice.

The carefully fashioned impression was that the GCCSI reflected global commitment, boasting over 300 members from all over the world, including dozens of national governments, a wide range of companies, international agencies and environmental groups. In reality, though, it has been and remains an Australian show, with Australian taxpayers providing some 99 per cent of the funding by some estimates. Given $100 million a year to spend, Rudd's GCCSI demonstrated an extraordinary capacity to burn money. Its Chief Executive would tell the *Sydney Morning Herald* of the big budget, 'It's actually impossible to spend that amount of money responsibly'.[92] Seventy-eight staff were hired, $400 000 a year was set aside to run the board, including first-class travel – a perk other lucky institute staff enjoyed en route to a host of lavish events it put on from Paris to Kyoto to promote itself, and a fast-growing pile of expensive reports prepared by consultants. These reports included $2.6 million for one on CCS projects in various stages of development around the world – costly confirmation that not one large-scale integrated CCS project in the world then operating involved coal-fired power.[93] With fiction perhaps more enjoyable than reality, the institute funded another $1.3 million report on an 'ideal portfolio' of CCS projects. There was also $1.13 million for ICF International for a series of workshops and a consultancy report on how to define a 'carbon capture ready' coal-fired power station. The CSIRO (working with the IEA) landed a three-year $4 million contract to analyse public awareness of CCS in Australia and elsewhere, and work so that CCS promoters could more effectively sell their messages.[94] In its first two years the institute ran up a staggering $54 million in operational expenses and another $13 million on 'enablers'.

In practice, this involved lots of big cheques for organisations to use Australian taxpayer dollars to run events, attend meetings, give speeches and write blogs, to create the illusion of impressive momentum. For example, when the IEA initiated its Greenhouse Gas Research and Development Program 'for regulatory work' promoting CCS, it did so with GCCSI funding. Another example involved the

IEA's CIAB workshop in New York on 'CCS: bridging the commercial gap' (perhaps predictably, it recommended stronger government support and more government funding). To encourage CCS boosting from different directions, the institute gave the IEA $20 million, the Asian Development Bank $21 million, and the Clinton Foundation $10 million. Internal documents reinforce that the GCCSI, though government-funded, saw itself as a CCS activist group. A draft 2010 PR strategy explained that if the public and governments acknowledged the hurdles for CCS and legislation was 'enacted targeting the destruction of these barriers', their strategy would be a success. This was a government propaganda house dressed up as something very different, and handed to the coal industry to run without so much as a public tender; an agency whose sole purpose was to sustain the impression that CCS was feasible.

Not surprisingly, the GCCSI was keen to build the appearance that some environmentalists were behind CCS, so more cheques were written to sympathetic green groups. An undisclosed amount was provided to The Climate Group, which among other things hosted public lectures, wrote positive blogs on the GCCSI website, and prepared an upbeat report on the readiness of private finance to support CCS (badged with the breezy slogan 'CCS is part of THE CLEAN REVOLUTION').[95] The Institute was a platinum sponsor of the Climate Group's Climate Week in New York City.

In Norway, with €900 000 from the GCCSI, a group called Bellona embarked on a 2.5 year program to foster support for CCS, to 'promote new ideas and overcome barriers'.[96] Australian taxpayers appear to have funded eight new staff at this green group half-way round the world, including an 'advocacy trainer', 'political and technological experts' to develop and lobby for country specific CCS 'roadmaps', and two PR officers to run and expand the group's CCS website (which features a world map giving the misleading impression that CCS is booming around the globe). The culmination of the GCCSI's efforts to recruit industry-friendly environment groups willing to take a pro-CCS posture has been establishing the Environmental NGO Network on Carbon Capture and Sequestration (ENGO Network on CCS). Among others, it includes the World Resources Institute, the Natural Resources Defence Council, and the Sydney-based Climate Institute. With an undisclosed amount of financial support from the GCCSI funding for pro-CCS reports, this network now proclaims that CCS 'should be considered a critical mitigation technology'. Seemingly oblivious, the Climate Institute declares 'CCS technology is available today, enabling the deployment of the technology to begin worldwide immediately'.[97]

Perhaps even more galling for Australian taxpayers than watching their governments blow hundreds of millions of dollars on failed CCS projects at home, and bankrolling PR to create an illusion of international momentum, has been the tens of millions of dollars wasted by the GCCSI on failed projects outside Australia. On the basis that they might accelerate the deployment of CCS globally, $50 million was allocated annually to directly subsidise projects outside Australia.[98] Most of the funding has gone to projects which at the time of writing are either on ice or have been cancelled.[99] These include the shelved Keephills project in Alberta, Canada, which was given $5 million, and the stalled Tenaska Trailblazer project in Texas, which received $7.7 million. As the institute's extraordinary capacity for waste becomes clearer, and as more projects that it has funded fail, there has been a withdrawal from directly funding CCS internationally. However, the GCCSI-funded global PR show rolls on, limited only by the impending expiry of the funds allocated to the organisation.

The zombie won't quit

As it gets clearer that CCS will take longer, and cost more than expected, more people are seeing through the spin and illusion of progress. Though CCS may be technically feasible and commercially viable one day, it won't be deployed on the scale required to rein in coal emissions in any meaningful timeframe. When that reality is factored into the CCS narrative, continued coal use is only consistent with the most severe climate change scenarios. After a decade bearing so few results, we might reasonably expect a broad backing away from the clean coal story in general and CCS in particular. Yet the zombie dances on as people with no back-up script chant the mantra they know. Pockets of unfathomable optimism remain. One lobbyist for the Victorian division of the MCA recently proclaimed that progressing clean coal technology could happen 'in the blink of an eye'. However, much of the upbeat rhetoric has become more sanguine to suit the circumstances. Now, we're told, failures were always anticipated, that this was always a long haul. The NSW Minerals Council, representing many Hunter and Illawarra coal companies, says now that the collapse of projects 'is not unexpected', and that they 'should not be taken to reflect a failure of the technology itself'. The ESAA, representing Australia's predominately coal- and gas-fired power generating companies, explains that 'we need to get used to failure as part of the journey – many technologies will fail before some succeed'. Meanwhile, runaway growth in coal use without CCS isn't viewed as a reason to finally question coal use, but as something that makes CCS more essential.

Political droids also march on. Inertia and the lack of a viable alternative leave CCS the alibi of choice for now. In 2011, even as his government was slashing funding for CCS, Climate Change Minister Greg Combet told a Hunter Business Chamber audience: 'The Government believes that low emissions coal technologies, including carbon capture and storage, are an essential component of the transition to a carbon-constrained economy'.[100] The ministerial press releases have changed though. Now it's all about government-run, rather than privately conceived projects – the South West Hub in WA and CarbonNet in Victoria. Only these are not real projects, but dreamy bureaucratic envisions of CO_2 pipelines all coming together in a handy location, if and when the idea ever stacks up. Not that these fantasies make much difference if realised; the emissions they might save would be wiped out at least 25 times over by new coal mines proposed in the Galilee Basin alone.[101]

Not surprisingly under the circumstances, the definition of clean coal is in the process of evolving to survive. In some ways, it's simply a case of back to the future, as Big Coal returns to the pre-CCS era. The idea of burning coal more efficiently is enjoying a renaissance – once again badged as clean coal. Various types of coal-generating technologies are indeed more efficient – from the Integrated Gasification Combined Cycle (IGCC), which burns coal more efficiently by gasifying it first, and then uses the excess heat to drive a steam turbine (hence combined cycle); to the combustion of pulverised coal under high pressure and at extreme temperatures – otherwise known as 'supercritical' or 'ultrasupercritical'. These technologies generate the same amount of electricity with up to 40 per cent less coal – hence lower emissions. So, as long as the focus is on the emissions per kilowatt hour of electricity produced, rather than the spiralling quantity of emissions from coal burning overall, a surprising number of people are still impressed with the idea that replacing old power stations with more efficient new ones can adequately clean up coal while we wait a few decades for CCS to become viable.

It makes a nice story, except that most new coal-fired power isn't replacing old coal-fired power – it's in addition – and while there are more efficient coal technologies around, most coal-fired power being added is the cheaper, less efficient 'subcritical' variety that has contributed so much to the problem already. As the IEA notes, 'subcritical technology has continued to dominate recent build'.[102] Most worryingly, it's a trend replicated in developing countries where most new coal-fired power is being added. In China, nearly twice as many old-style subcritical plants were added in the last decade as the more efficient alternatives.[103] In India, it's even worse – almost all new capacity has been subcritical. Hence, there has been almost no gain in the efficiency of the global coal-fired electricity fleet in spite of all the talk of more efficient coal use. According to the IEA, 'Worldwide, coal-fired power plant efficiency averaged 35.1 per cent in 2007, compared with 33.5 per cent in 1971'.[104] As a consequence, global CO_2 emissions from coal combustion have increased around 60 per cent since the year 2000.[105]

Stuck with an implausible 'efficiency gains will fix it' argument, and CCS decades away from viability, the clean coal pitch risks a complete collapse of public support. And it is for this reason that Big Coal is keenly broadening the definition of clean coal to buy even more time. To that end it now promotes a dazzling array of further ways to reduce the footprint of coal burning. One option is 'co-firing', which in practice means throwing some biomass (e.g. forestry/agricultural waste) onto the coal pyre. The biomass can be sold as 'renewable' and 'carbon-neutral', and the overall emissions per kilowatt hour of electricity produced can be said to be lower. Similarly, solar or wind power arrays are being installed next door to coal-fired power stations, enabling the company to claim a slight improvement in the overall emissions per tonne of coal burned, and declare their operation a 'hybrid' power station. The renewable components of the operation tend to be preposterously small – in Queensland, for example, the Kogan Creek Solar Boost project is heavily promoted as 'the largest solar thermal project in the southern hemisphere', its owner, CS Energy, claiming that 'Projects like this are essential for CS Energy to successfully transition to a low carbon future'. In practice, however, the solar add-on will provide less than 1 per cent of the electricity produced by the power station.[106] The spin is remarkably solar powered, but not the power station.

Along with 'hybrid-coal' power stations, Big Coal is promoting a host of other clean coal options. These include everything from storing CO_2 in mineral form for use in concrete to using CO_2 to boost the production of urea, to using it in the production of plastics. Though there's little evidence to suggest these options can lead to an overall reduction in coal emissions, the industry gives the impression that the vast majority could well be viable in the next ten or so years. In PR terms, one technology is shaping up as the coal industry's next silver bullet – the short- to medium-term alternative to CCS. Clean coal advocates are latching onto the idea of 'recycling' CO_2 by pumping it into ponds adjacent to power stations and steel mills to turbo-charge the growth of algae. This algae can then be turned into diesel or even aviation fuels and other products. Supporters point to a cornucopia of valuable by-products that could be made from algae – fertilisers, feed for cattle and fish farms, vitamin supplements for humans and soil-enriching biochar. The concept is sold as a big part of the solution to energy and food security, and as 'renewable'. Compared with piping a waste product thousands of kilometres away and hoping it stays buried, this is 'carbon capture and recycling' (CCR) – the PR spruik is infinitely easier.

The déjà vu parallels with the CCS push are as uncanny as they are unnerving. Having just been thoroughly taken for a ride by upbeat CCS merchants, politicians from both sides are lining up to support the idea with an astonishing technological love-struck naiveté. The public purse is again being opened to support pilot-scale demonstration plants adjacent to large coal-fired power stations along Australia's eastern seaboard.[107] Mining companies and coal-fired power producers are keenly tipping funds into companies that promote algae, not just as the solution to CO_2 emissions from coal, but as that path to independence from Middle Eastern oil, and as the only way to clean up global aviation. Another group of scientists used to working in obscurity suddenly find themselves on the edge of a financial bonanza. Again there's bold talk about what's possible. One prominent algae technology proponent, MBD Energy, plans to build a 'commercial pilot plant' at large coal-fired power stations in Queensland, NSW and Victoria.[108]

The political rhetoric doesn't just mirror what we heard about CCS a decade ago – we see the same people in the mirror – Ian Macfarlane calls CO_2 reuse via algae 'probably the most promising (technology) for capturing carbon'; Greg Hunt says it may be 'one of the most viable technologies'.[109] Tony Abbott is all for 'putting emissions to good use rather than just burying it,'[110] and says the technology could cut emissions from our largest coal-fired power stations by at least 50 per cent within a decade. Robin Batterham joined the board of MBD Energy in 2012. The Gillard government has given carbon capture and recycling CCR access to subsidies available under the Clean Energy Finance Corporation withheld so far from CCS, and support from Labor politicians is as upbeat as from the Coalition. Peter Beattie (now paid $1000 a day by the Gillard government in the part-time role of 'Resources Sector Supplier Envoy') is glowing about the algae business. Once again, he sees the technology not only as the solution to CO_2 from coal, but as a golden opportunity to build a new export industry.[111] Same old script, only potentially more potent because this is all about recycling CO_2 into a host of desirable products rather than treating it as waste that needs storing.

As with CCS though, this is another PR-based mirage. For starters, the process is not as green as it initially sounds. To produce the CO_2 being 'recycled', large quantities of coal are burnt, so the process is not 'renewable' as it relies on burning non-renewable coal. The CO_2 from the coal still ends up being added to the atmosphere rather than staying underground. The process also suffers from many of the same problems as CCS. Most obviously, there's the question of financial viability. Almost all the projects being implemented worldwide are highly visible, small-scale pilot projects, not viable without taxpayer support. Here again, the hope is that carbon prices will drive large-scale demand for CO_2 recycling, but like CCS, CCR won't be viable in the foreseeable future without public subsidy. By some estimates, algal-based biofuel isn't viable until oil prices reach US$800 a barrel, and the carbon price alone is unlikely to justify it before 2020, or perhaps much later.[112]

Beyond viability, there are other problems reminiscent of CCS – the algae-based fuel can displace some diesel, but emission savings are eroded by things glossed over by the spruikers, such as the energy used to keep the water moving in the algal ponds, the energy and resources used to produce the fertilisers used to feed the algae, which don't merely survive on CO_2 and the sun. Then there's the problem of space – in many cases there's not enough land adjacent to the coal-burning facility to accommodate the number of algal ponds required, so the CO_2 would have to be piped great distances, requiring more energy still. Last, there's the telling hassle that the process doesn't work without sun, so at most, around half of the emissions of a power station could be captured and recycled, although the US Department of Energy says 20–30 per cent is more realistic.[113] So, the algae solution won't be rolled out soon enough, and even if it was it couldn't cut total emissions from coal globally. This leaves coal-fired algae's main practical impact as PR – another tool to justify the continued expansion of coal use.

Given that, perhaps we should take stock before embracing a new round of clean coal optimism. The unpleasant reality is that after 30 years of clean coal talk, the emissions from coal have more than doubled.[114] We can choose to ignore this history and run the risk of repeating it, or we can heed it. The lesson from what we've seen is pretty clearly that coal can't be clean, no matter how hard the industry tries to persuade us otherwise, and no matter how much we might want to believe them. It's time to move beyond it.

7 OVERTHROWING BIG COAL

> '... as to energy policy's polluting past, it can be summed up in one word: coal. And even though the coal industry doesn't totally know it yet or is ready to admit it, its day is done. It used to be said that 'coal is king', and regrettably coal continues to be king [in some parts of the world] ... but here in the US, I'm happy to say, the king is dead. Coal is a dead man walking.'
> MIKE BLOOMBERG, Mayor of New York[1]

The chipped china collection of Australian animals in the cupped hands of the 12-year-old girl was just about all that remained of her worldly possessions. The rest had gone up in flames as a massive bushfire raged across the Tasman Peninsula in south-east Tasmania in early January 2013. The girl's family home – along with another 110 nearby – had all been destroyed within a few hours. Eight kilometres up the road in Dunalley, one family had a harrowing escape: two grandparents clutched their five grandchildren in chest-deep water beneath a jetty while the fire raged through the surrounding bush. The fire which swept through that day occurred on what was the hottest day in the hottest month on record in the state. The weather conditions were so extreme that the fire was classed as having a 'catastrophic' fire danger rating, a category only created in recent years.

In the summer of 2013, 123 Australian weather records were broken, mostly for high temperatures. There were hundreds of fires in the south-eastern states of Australia and numerous localised, 'catastrophic' alert levels. Fires burned down homes and tens of thousands of hectares of farmlands and forest, and in Victoria a man was burnt to death in his car. If the bushfires weren't bad enough, the remnants of Tropical Cyclone Oswald swept through Queeensland and down the NSW coast. In Bundaberg 3000 homes were inundated, many damaged beyond repair, and over 7500 people stranded. In NSW over 20 000 people in 39 communities were isolated by floodwaters and two tragically drowned. The federal government's Climate Commission, which was established to provide public information on climate change, dubbed it 'an angry summer', noting that globally there has been a ten-fold increase in extreme weather events since the 1950s.[2]

While no one event can be neatly labelled as being caused by global warming, the record-breaking weather events are entirely consistent with the changes expected as the climate alters. A few years before the fires, University of Melbourne Professor of Meteorology David Karoly bluntly summarised the implications of what a 4-degree warmer world[3] would mean for bushfires, 'We are unleashing hell on Australia'.[4]

A few months earlier, a review of the latest scientific research commissioned by the World Bank bluntly warned that greenhouse gas emissions are now growing so fast that the world is on track for a 6-degree increase in global average temperatures, of which 4 degrees could occur by 2100, or possibly even as early as the 2070s.[5] The report by the Potsdam Institute for Climate Impact Research and Climate Analytics, a leading research centre on climate impacts, noted that the oceans had grown warmer and more acidic, and ice loss from the Greenland and Antarctic ice sheets is accelerating.⁶ The World Bank's President, Dr Jim Yong Kim, summarised that a world that was 4 degrees warmer was likely to cause:

> the inundation of coastal cities; increasing risks for food production potentially leading to higher malnutrition rates; many dry regions becoming dryer, wet regions wetter; unprecedented heat waves in many regions, especially in the tropics; substantially exacerbated water scarcity in many regions; increased frequency of high-intensity tropical cyclones; and irreversible loss of biodiversity, including coral reef systems.⁷

The IEA, traditionally a stolid documenter of global energy trends, has become alarmed too, warning that even if current commitments by national governments are honoured, it will be too little, too late to meet agreed targets.⁸ Corporate advisers PriceWaterhouseCoopers warned that the rate of 'decarbonisation' is one-quarter of what is needed.⁹ The IEA, World Bank and PriceWaterhouseCoopers all agree that global electricity supply – which creates almost 40 per cent of the global greenhouse gas emissions – must be 'decarbonised' if the world is to have any chance of limiting temperature increases to the internationally agreed target of 2 degrees, let alone the 1.5 degrees leading climate scientists argue is necessary.¹⁰ Even Greg Sullivan, Deputy Chief Executive of the Australian Coal Association, agrees that the world is on track for at least a 4-degree temperature increase.¹¹

Extreme climate events may be upon us, but key political players from Australia's largest coal-mining and coal-burning states remain steadfastly disengaged. Less than two years after Victoria's devastating 2009 bushfires, the state's then Treasurer, Kim Wells, was fixated on defending its brown-coal-fired power stations: '… we want to protect that competitive advantage come hell or high water.'¹² Long-term climate scenarios suggest that Victoria will get both.

At the time of the January 2013 floods, which swamped parts of northern NSW towns, Premier Barry O'Farrell was equally emphatic, 'Let's not turn this near disaster ... into some politically correct debate about climate change. Give me a break!'.[13] Following the devastating Queensland floods of 2011, former Premier Peter Beattie acknowledged that the floods, along with the preceding drought, 'were climate change in action'. However, he thought the policy priority was clear: 'hopefully, Queensland can resume coal exports quickly to protect its markets'.[14]

Criticism by a UN team of plans for major new coal export ports potentially affecting the Great Barrier Reef World Heritage Area prompted current Queensland Premier Campbell Newman to proclaim that 'we are not going to see the economic future of Queensland shut down ... We are in the coal business. If you want decent hospitals, schools and police on the beat we all need to understand that.'[15] It is a false framing of the climate challenge but one that has been peddled by the coal industry for so long that it is now accepted as an article of faith by most politicians, both federally and in the key coal states.

To all but coal's cheerleaders it is crystal clear that the expansion or maintenance of global coal consumption is incompatible with ensuring a stable climate, let alone protecting air, land and water quality. The failure of international climate negotiations to effectively respond to the escalating crisis has created a level of pessimism that there will ever be agreement among the world's governments to curb fossil fuel burning.

If international climate negotiations are going nowhere fast, how can Big Coal be overthrown? Across the country, farmers are opposing the destruction of farming land for new coal and coal seam gas mines, residents are opposing air pollution from coal trains and new export terminals, and the glory days of the old coal-burning power stations are fading fast. Tired of hoping that governments will take the lead, a new movement consisting of farmers, residents and environmentalists is emerging and challenging the industry as never before. The souring appetite for coal in many countries has slowed the rate of new projects from boom-time mania to a far slower pace. Banks and other financial institutions – especially in the wake of the global financial crisis – are increasingly wary of big coal projects.

Around the world a similar anti-coal rebellion is flourishing and having an impact. In the United States the use of coal for power generation is in decline domestically, while in most of Europe the retirement of old coal-fired power stations is set to accelerate in the next few years. In China, the terrible effects of air pollution emitted from burning vast volumes of coal has stirred up a remarkable grassroots political rebellion, causing it to teeter on the brink of capping and then reducing coal consumption within a few years.[16] The old coal industry narrative, that demand for coal is only going up and will for ever more, is looking decidedly out-of-date.

If political opposition to the planned expansion of coal production becomes much more powerful in the key exporting countries that are supplying the Asian and Indian markets, the ripple effects will be profound. If one or all of Australia, Indonesia, the United States and Canada slowed or cut coal exports, those weighing up the risks associated with 30–50-year investments in new coal power stations in Asia and India would be far more likely to play safe and opt for alternatives.

If China and/or India change tack in the next few years from wholeheartedly wanting more coal for power stations and steel mills to putting greater emphasis on cleaner alternatives such as renewable energy, the effect on the demand for new coal mines will be massive. Considered unthinkable a few years ago, most analysts now seriously contemplate a scenario of Chinese coal demand slowing, peaking and then declining. Some consider it could happen within a few years, others in a decade or more. Whether China and India pull back from embracing coal hinges on what weight is placed on protecting the health of the public in the coal-burning cities, the availability of water, the availability of finance for coal plants and the economics of renewable energy.

Coal demand would drop even further if the biggest coal-consuming countries shift to retiring old coal plants and replacing them with renewables and reducing demand through energy efficiency programs. Reduced coal demand would stall proposals for new mines and infrastructure in supplier countries, while lower returns would make finance harder to obtain.

An equally powerful effect would be created if the banks and financial institutions that have profited so handsomely from Big Coal were to pull back from financing new power stations, mines, ports and other infrastructure. If global multilateral agencies such as the World Bank were to chart a new direction, they would have a powerful effect on the other multilateral banks and the private finance industry.

Few think that the currently mired international negotiations will make substantial progress any time soon. However, the global stalemate has prompted the proposal of a parallel international path dubbed 'Cooperative Decarbonisation'. At its heart is the need to build a coalition of countries willing to dramatically cut emissions, recognise unburned fossil fuels as global liabilities, protect remaining forests as carbon stores and make renewable energy universally available. Participating countries would cooperate in adopting emissions reductions strategies or, where cooperation is lacking, take unilateral action. Ultimately it is hoped that a bottom-up approach would eventually lead to a more effective climate-centred international agreement.[17] Similar style coalitions – some involving or led by Australia – have been effective in relation to issues such as banning landmines, curtailing proposals to mine Antarctica and smoothing the transition to democracy in Cambodia.

As a country we face a choice: to either pursue a coal-centred strategy until either the global community or the coal market dictates otherwise, or to chart a course for coal-free prosperity and embrace all the opportunities and challenges that presents. Decarbonising Australia – a journey we have already half-heartedly begun – will allow us to head in a new economic direction: one that reduces household and business electricity costs through the efficient use of electricity and energy, and stimulates the widespread deployment of renewable energy.

Such an approach would eliminate the ongoing economic liabilities that come from coal power: the permanent removal and destruction of land from farming and other industries, the health impacts of air pollution, the pollution of waterways, the long-term costs of greenhouse gas emissions and the ongoing social conflict. Hand-in-hand with domestic decarbonisation there needs to be a coal exports phase-out aimed at creating a more resilient and diversified economy. In this way, Australia can avoid the problems created as regional and state economies make rapid shifts in a volatile global market by a commodity that has all the long-term desirability of asbestos mining. Australia could be a significant leader of an international decarbonising coalition; especially as we are such a big coal burner and exporter – but only if we practise what we preach.

A domestic makeover

With one of the highest levels of both per capita consumption of electricity and of electricity sourced from coal-fired power stations in the world, Australia has substantial scope to rapidly shift to being a more energy efficient country with electricity supply drawn from renewable energy. However, there is a sobering backdrop: since 1990, when Australia first began canvassing greenhouse gas cuts, emissions from the electricity industry have increased by 47 per cent.[18]

All up, Australia now has a fleet of 23 coal-fired power stations, which together emit over one-third of all Australia's greenhouse gas emissions.[19]

With many existing coal power stations entering old age, the prospect of making the transition to coal-free domestic electricity is nowhere near as daunting as the coal industry makes out. No new coal-fired power stations are likely to be proposed, given currently falling demand for electricity, public opposition and deep-seated wariness of the big banks. The introduction of even just a low carbon price has tipped the scales even further against coal-derived electricity, even though generous cash handouts, free pollution permits and access to cheap carbon credits from overseas for big carbon emitters have slowed their demise. Rapid increases in power prices have spurred householders and businesses to embrace energy efficiency measures, such as less power-hungry heating, cooling, lighting and appliances.[20] The federal government's 2009 household insulation program has had an effect too. Plummeting solar power costs, government subsidies and feed-in tariffs have prompted over 1 million households to install solar panels. The high Australian dollar has forced the restructuring of export-oriented industries, further undermining electricity demand. Since the end of 2008, when electricity generation peaked, greenhouse gas emissions from electricity have dropped by over 11 per cent due to falling total demand and increased generation from renewable energy sources and gas.[21]

Around the country coal-fired power stations are being shut down, mothballed or run less frequently. In 2012 two coal-fired power stations – the Munmorah Power Station in NSW and the Swanbank B Power Station in Queensland – were permanently closed down.[22] Half the capacity at the Tarong Power Station in Queensland has been closed until at least until late 2014,[23] while half the capacity of the Wallerawang Power Station in NSW has been shuttered until early 2014.[24] The decrepit Energy Brix power station near Morwell in Victoria is scheduled to close by mid-2014 at the latest.[25] Other coal power stations have been downgraded: the Northern Power Station, near Port Augusta in South Australia, is on limited duties until October 2014, while the nearby Playford B Power Station has been relegated to summer-time duties.[26] In WA, the Muja Power Station will be run significantly less.[27] It is an unprecedented and rapid move away from coal-fired power stations, demonstrating how quickly a shift away from coal can be made. Proposed gas-fired power stations have also been shelved.[28]

Domestic coal power stations are also being propped up by massive fossil fuel subsidies from taxpayers and electricity consumers. While Australia has vowed that as a member of G20 group of countries it will remove fossil fuel subsidies – estimated at between $9 and $12 billion – it is avoiding making cuts by playing semantic games over the definition of a subsidy. Instead of embracing the 'polluter pays' principle, the government has agreed to put $5.5 billion from the carbon tax into the purses of the biggest power generators in order to ensure 'energy security'. Electricity consumers have provided billions of dollars more over the last 30 years to a handful of aluminium smelters,[29] which account for approximately 13 per cent of all Australian electricity consumption. Even with all the subsidies, the industry is more insecure than ever as the new low-cost centre of global aluminium production shifts to countries such as China. The domestic industry is demanding and getting even more public subsidies.[30]

Even with the subsidies, the Bureau of Resources and Energy Economics (BREE) forecasts there will be many more coal power station closures. They estimate there will be no brown coal power stations operating at all in Victoria within a decade or two, and that, at best, by 2050 black coal power stations will produce less than half of what they produce today.[31] For all the hype from the coal lobby about coal power being here forever, Australia is already well on the way to ending its love affair with coal power.[32]

Embracing a shift to renewable energy fundamentally changes the geographic distribution of economic benefits away from where coal deposits lie to where the sun shines – which is just about everywhere in Australia – and where the wind blows best. Wind power and other renewable electricity exports from Tasmania and SA are already significant interstate income earners. A coal power phase-out would have no impact in the Northern Territory, which doesn't mine or burn coal for electricity and is not connected to the national grid. Tasmania and the Australian Capital Territory have no coal power stations, although they both import coal power from the national grid to differing degrees. The main local impacts of a coal power phase-out would be confined to Victoria's Latrobe Valley, the Hunter Valley in NSW, Central–Northern Queensland and to a lesser extent, the Collie area in WA.

The expansion of renewable electricity on the supply side needs to be complemented by increased efficiency on the demand side. Higher energy efficiency standards for appliances, the adoption of far more efficient lighting, the accelerated deployment of solar hot water, higher household insulation standards – all proven existing cost-effective technologies – would combine to help drive down per capita household electricity consumption. Most industry can also make substantial efficiency gains, whether from building upgrades, the installation of more efficient plant and equipment or changes to industrial processes. Greater energy efficiency not only boosts profits from business, but increases disposable household income with flow-on economic benefits. A reduction of 10–30 per cent in electricity demand – the common range considered feasible from end-use energy efficiency measures – alone could account for the closure of several coal-fired power stations.[33] However, there is one hurdle: the more governments allow cheap carbon credits from overseas as counting towards domestic carbon reduction targets, the weaker the domestic incentive for domestic energy efficiency.

Faced with falling domestic electricity demand and increased renewable energy supply, the incumbent fossil fuel power generators are frantically lobbying like old-style protectionists to shore up their fading empires. They have successfully lobbied to slash feed-in tariffs for solar power in most states, and are pressing to weaken or axe both the carbon price and the 20 per cent Renewable Energy Target, while retaining existing fossil fuel subsidies. There are also moves by climate change sceptics and their supporters to curtail the growth of windfarms through tough planning restrictions and restrictive regulatory standards.

While slashing the amount of domestic coal burned is within reach, Australia's biggest contribution to greenhouse gas emissions comes from coal exports, which BREE estimates could reach 585 million tonnes by 2050, or almost 15 times as much as is projected to be burned in Australia.

Coal's clouded future

Marcus Randolph, the head of BHP Billiton's coal division, could hardly be accused of being an anti-coal activist. However, in late 2012 he acknowledged that the long-term prognosis for thermal coal over the next 30 to 40 years was 'very clouded' and that 'I suspect, the usage of thermal coal is going to decline.'[34] It was a telling comment, as state and federal government are placing their faith in huge growth in demand for coal for power stations to spur a rash of mega-mines in NSW and Queensland.

According to the IEA, there is 'much uncertainty for international steam coal markets and prices'. In part, this is because if the global community is serious about limiting the global temperature increase to 2 degrees, coal demand would have to drop by almost an estimated 1.8 billion tonnes of coal a year by 2035.[35] It would require even more dramatic cuts to reach the less dangerous 1.5-degree temperature increase target. Estimates of future coal use hinge on three key factors: what replaces the soon-to-be exhausted existing coal-fired power stations, which produce as much as a third of the world's electricity; how China and India will cater for new electricity demand; and whether the international community takes global warming seriously. For its part, when BREE forecast that Australian exports would almost double it assumed Chinese coal consumption would not plateau until some time in 'the latter half' of the forecast period to 2050.[36]

The staggering Chinese air pollution crisis of January 2013, just one month after BREE published its forecast, may well undermine the coal industry's grand expansion plans. Coal and car pollution sent small particle pollution levels in Beijing to over 35 times higher than that defined as safe by the World Health Organisation (WHO). In the short term, high levels of PM2.5 particles can cause breathing difficulties, high blood pressure and heart attacks. Longer term exposure can cause lung cancer and dramatically reduced life expectancy. As the smog persisted, public alarm escalated. Hospital admissions soared and government agencies warned the public to stay indoors. According to the Chinese government's Ministry of Environmental Protection, over 600 million people across the country were affected by the toxic smog, which also spread to Japan.

Hundreds of millions of Chinese may have been exposed to the putrid air, but the following month the NSW Minerals Council was in a celebratory mood: 'NSW coal exports to China grew by 210% from 2010–11 to 2011–12', it tweeted.[37] The ACA approvingly echoed the message.[38] However, the coal industry's optimism may be short-lived.

The smog crisis has pushed the nervous Chinese government to act. The government's previous PM2.5 standard of 75 micrograms per cubic metre, three times the WHO standard, was scrapped and a level of 35 micrograms per cubic metre by 2030 adopted. Underlying public and government concern is recognition that without drastic action the country's PM2.5 pollution could increase by up to 70 per cent. The government has also restated its commitment to cap coal usage, albeit at slightly higher than current levels, and announced plans for a carbon tax.[39] While BREE[40] and some commentators have expressed doubt that the government will stick to its targets,[41] others disagree. As a Deutsche Bank analyst based in Hong Kong noted, 'Can anyone imagine what would happen when it [PM2.5] rises to 1,500?'[42] Ma Jun, the founder of the Institute of Public and Environmental Affairs, a Beijing-based environmental group, explained that the government '[doesn't] want this issue to get out of control as it could become a major source of social unrest'.[43]

In an outline of how to reach the new PM2.5 target, Deutsche Bank analysts suggested that, even allowing for the installation of pollution control equipment on newer coal plants, the growth in coal demand would need to be cut in half until it peaked in 2017 and then total consumption cut by one-fifth by 2030.[44] If Chinese coal demand increases for just a few more years, plateaus and then falls rapidly, the implications for the global export coal industry would be profound. China would switch from importing almost one-quarter of the global market to importing nothing. Or it could even become an exporter again. With an oversupplied global market, prices would fall, funding for new mines and infrastructure evaporate and high-cost mines would be shuttered. A dramatic cut to coal-fired generation capacity would be matched by a dramatic further expansion in renewables such as solar and wind,[45] with an inevitable cost-reduction spillover effect on the global market. While there is considerable doubt about if, when and by how much China could cut coal consumption, the uncertainty alone casts a huge shadow across the export coal market.

The Chinese government has also signalled its intent to crack down on pollution from its huge steel-making sector, which will have a ripple effect on the metallurgical coal industry.[46] Australia is the world's largest metallurgical coal exporter, while China is the world's second largest importer, after Japan. Despite the coal industry's claims that coal is a key component of steel manufacturing, alternatives exist. Approximately 30 per cent of the world's steel is produced in electric arc furnaces, which rely on scrap steel and can be powered from renewable electricity.[47] While some electric arc furnaces use a small amount of coal, in 2011 over 55 million tonnes of steel was produced from gas-fired direct reduced iron plants,[48] with millions of tonnes more capacity planned at plants from Iran to the United States. Other technologies exist and are being implemented to reduce the coal intensity of steel production in both electric arc furnaces and conventional blast furnaces.[49] If the global atmosphere was no longer treated as a free dumping ground for coal's emissions, then there would be more incentive to shift production to coal-free alternatives and substitute other materials for steel in end uses.

If China slows or reverses its coal-burning binge, the fate of the remnants of the Australian coal boom would then largely hinge on what happens in India. At face value the impact of the projected power needs of India are staggering. Over 500 coal-fired power stations have been proposed, amounting to 608 gigawatts of capacity — 12 times larger than Australia's entire electricity grid.[50] Look a little deeper though and a familiar tale emerges. Hundreds of projects are speculative or, at best, uncertain, as they lack one or all of government approvals, land, coal, water or finance. Many of the projects are clustered around a handful of cities and it is clear that not all could proceed, given the size of the market.[51]

The Indian government, stung by major blackouts in early 2012, has buckled to coal lobby pressure and is seeking to increase domestic coal mining and new coal-fired power stations. Despite this, there is strong resistance to many proposed new coal plants — villagers opposing the acquisition of agricultural land and the allocation of vast amounts of water. Pollution from existing plants is already so bad that the IMF has estimated coal plants cause approximately 70 000 premature deaths every year in India,[52] with other estimates putting it even higher.[53]

Some of the new private-sector power projects, such as the Tata Mundra Ultra Mega Power Project – the first of up to ten 4000 megawatt 'ultra-mega' coal-fired power stations promoted by the Indian government – have encountered major financial problems.[54] Other private power companies are struggling too. In February 2012 the Reserve Bank of India went so far as to caution Indian banks that it wouldn't bail them out if new coal plant loans went sour.[55] Even though the World Bank has acknowledged that climate change will compromise the development opportunities of the world's poorest countries, it remains a funder of coal projects, along with many other public finance institutions.[56] However, if the World Bank changes tack to focus on renewable energy and efficiency instead of coal, the already precarious economics of the coal power station boom in India become even more uncertain.

If the demand side of the equation is looking uncertain, what about the supply side? Faced with a downturn in domestic demand for coal in the United States, some coal companies have floated plans to build new export coal ports on the west coast. Despite grand plans of cashing in on the Asian coal market, some of these projects have faltered, while others are encountering strong opposition from local groups and the Sierra Club's Beyond Coal campaign. Without new coal infrastructure, US coal companies will have little scope for exports of low value thermal coal beyond a minor amount through the existing export terminal in Vancouver, Canada.[57]

This would leave Indonesia and Australia as the main suppliers of choice in the Asian neighbourhood, especially for power station coal. While coal exports from Indonesia have grown phenomenally in the last decade, there are limits. Two years ago, the Indonesian government decided that coal could only be exported if it was sold at a price benchmarked against the global market price. This sent shockwaves through those Indian power producers that had built power stations predicated on imported Indonesian coal bought for approximately US$35 a tonne, rather than two or three times that price. The Indonesian government has also imposed restrictions on foreign ownership of its mines, making investors and financial institutions nervous. There are other pressures too: some coal deposits are in protected forests, while others are opposed by local landowners.[58] The government also remains concerned about exporting coal that it might want in the future for its own power stations. Major new coal projects are likely to be higher-cost operations as they will require new inland railway routes to be built.[59] And lower quality Indonesian coal can't be substituted for higher quality Australian thermal coal either because the Japanese power stations have been designed to burn higher quality coal.[60]

An Australian commitment to phase out coal exports, to match our domestic phase-out, would signal to other exporters and burners alike that, beyond say a decade-long transition period, the use of coal should be considered as unacceptable in a world that has less damaging alternatives to supply electricity and make steel. Such a move would be noticed worldwide and raise international expectations that sooner or later more countries will follow suit. An Australian coal export phase-out announcement would also signal to coal project developers – especially in China and India – that coal supplies could be far more expensive or less secure in a rapidly changing political environment. Given the limited additional coal supplies that can be brought quickly to market, it could be a decisive factor in tipping decisions away from the building of new coal plants.

A commitment to a coal export phase out would also help change the social norm about the burning of coal in the same way that banning indoor smoking was an important turning point in curtailing tobacco consumption. If smokers are allowed to smoke everywhere, more people smoke. If Australia were to move away from coal it would further isolate those banks and financial institutions that are considering funding new projects and leave the remaining large coal exporters and their supporters resembling a huddle of smokers banished from even being near the doors of buildings.

Is there life beyond coal exports?

In late 2012 BHP Mitsubishi Alliance (BMA) – a joint venture between BHP Billiton and Mitsubishi Development – announced that both its Norwich Park and the nearby Gregory mines in Queensland would close for good. The mines, the company stated, were victims of 'falling prices, high costs and a strong Australian dollar'[61] and, in the case of Norwich Park, the 2011 Queensland floods.[62] A few months later, Rio Tinto announced that, with the coal seams at the Blair Athol mine 'largely mined out', what was once the largest thermal coal export mine in Australia would close.[63] It was much the same story at Whitehaven Coal's Sunnyside mine in the Hunter Valley, when in late 2012 the mine was mothballed, with the company blaming low world coal prices.[64]

What happened at those four mines has been quietly happening right across the coal belts of Queensland and NSW. In the space of just nine months between May 2012 and early 2013, 15 700 coal mining jobs disappeared.[65] If mining industry employment multipliers are to be taken at face value,[66] over 105 000 direct and indirect jobs were lost. There was no coal industry advertising campaign lamenting the job losses or angry rallies on the steps of Parliament House. Nor was there even a consultant's report arguing how devastating the mine closures would be for the regional communities. For the most part there was just a matter of fact resignation that it is the reality of market forces. Families packed up and moved, some to new careers; others retired. Some of those who worked for BMA were given jobs at other nearby short-staffed company coal mines.[67] Most of the 140 people laid off from Rio Tinto's Blair Athol mine opted to negotiate a redundancy payout, while a few preferred deployment to other company mines.

The downturn in coal not only explodes the 'sky will fall in' mythology promoted by the industry, but also illustrates the challenges facing many of the over 120 mines in the export coal industry: high costs, the high Australian dollar, resource depletion and – by recent standards – low or at best volatile coal prices. Many of the existing mines are already precarious and most analysts expect thermal coal prices to remain stagnant for at least a few years. One analyst estimates that only the lowest cost new mines are now feasible.[68]

Whether existing mines are allowed to operate until the coal resource is exhausted, or are phased out early to protect the global climate, the implications of closure are the same: the need for new jobs for employees or assistance to reskill and relocate them where necessary, assistance to communities so they can diversify industries and adapt where possible, and the charting of an economic direction for a coal-free future by state and federal governments. The larger the coal industry grows, the harder a smooth transition becomes. The rapid growth of the mining industry has driven the Australian dollar higher. As a consequence, the rest of the economy (for example, construction and services) has been reshaped to service the export coal industry. Other more labour-intensive and sustainable sectors have been hollowed out as the fate of the economy is linked to how much coal-fired electricity is generated and old-style steel made in a handful of Asian countries.[69]

Having a more diverse and balanced economy would smooth out the ups and downs of the inevitable business cycle. Few coal boosters ever publicly air their doubts, although former Prime Minister Kevin Rudd noted that: 'We are in grave danger if mining runs into a wall, having shelled out the rest of the economy on the back of the high dollar ... We would be setting ourselves up for a mega-bust.'[70] Yet governments both state and federal, both Labor and Liberal, fall over themselves to encourage more coal mines, construct railways and ports. They fear Australia will 'miss out' on the high prices. The unstated assumption is that the quicker new mines are opened up, the better off Australians will be.[71]

It is a dubious assumption at best. Like a puffer fish bloating itself up to look larger and scarier than it really is, the coal industry likes to portray itself as a big jobs generator. In reality, it is a relative jobs minnow. By late 2012 the coal industry employed just over 45 700, in a national workforce of approximately 11 million. Only one in eight of those employed in the coal mining industry are women.[72] It employs only one-quarter the number of those working in the university and tertiary education sector, a few thousand more than the printing and publishing industry,[73] and is marginally larger than the Australian diary and milk processing industry.[74] As for the industry's claims that the proposed new coal projects would create new jobs and that an employment multiplier of three was reasonable, senior Treasury official, David Gruen told a Senate Estimates Hearing in February 2012:

> ... in a well-functioning economy like ours, with unemployment close to its lowest sustainable rate, it is not the case that individual industries can create jobs; they are simply redistributing them ... In a well-functioning economy that is achieving close to full employment, there really is not a multiplier ...[75]

Nor are past employment patterns much of a guide to future projects. In response to current low prices, Rio Tinto has not only vowed to get tough with its unionised coal workers, but it has foreshadowed extending its 'Mine of the Future' robotics program to its coal operations.[76] Driverless haul trucks, driverless trains, robotic drilling rigs, automated loaders and remote-controlled shipping operations are all part of its plans. In the WA Pilbara iron ore mine, where the robotics project was introduced in 2008, Rio Tinto boasted it had saved them from employing 900 more people.[77]

While the industry touts its credentials as a major employer in regional economies, most new jobs at the ever-more remote mines of Queensland are being planned for a FIFO workforce, or may end up being run from a control room in a city thousands of kilometres away.

A coal exports phase-out is likely to be far less disruptive for employees than feared as the mining industry has a notoriously high turnover rate among staff, with one estimate putting it at 26 per cent per year.[78] Many mining employees can and do shift from one commodity mine to another or do a short stint in a mine to earn big dollars, then shift to another industry. In one study, the Australia Institute examined data for employees in the mining sector over a period of seven years and found that those who left mining were not unemployed for long.

While boasting of jobs created by the industry, the ACA does not disclose the jobs destroyed by the industry. Occasionally, however, the costs of a new mine are revealed. Economic modelling, released in an Environmental Impact Statement prepared by Waratah Coal in support of the China First project in the Galilee Basin, explicitly concluded that the export of $4.6 billion of coal 'will likely place some upward pressure on Australia's exchange rate affecting the competitiveness' of the manufacturing industry. The company's economic consultants concluded that over an 18-year period from 2018 the project would cost 1600 jobs in the Queensland manufacturing sector alone, along with a host of other adverse economic and welfare impacts. The benefits from the mine, they concluded, would largely accrue to those who owned the mine or worked there.[79] Multiply those impacts by the number of mega-mines and we can see the high costs of a coal boom. On other occasions the conflict is less abstract, such as when Anglo Coal opposed the owner of a vineyard in the NSW Hunter Valley from establishing a $14 million tourism project comprising 23 tourist cabins and a wedding centre. Anglo Coal claimed the development would be 'unsuitable' because an extension it planned for its Drayton mine would exceed air standards by up to 20 per cent on a regular basis for 10 to 15 years. Instead of ensuring it complied with regulatory standards, Anglo sought to block the tourism project.[80] Such occurrences are being repeated right across the coal belt: coal mining companies squeezing out farmers, horse breeders, vineyards and residents, with an incalculable cost (see Chapter 2).

While the coal industry heralds the figure of approximately $40 billion in export earnings, it is an illusory benefit. If the Australian government levied a carbon tax at the same rate as it does for domestic emissions, over $20 billion would be taken off the headline figure straight away.[81] As most coal exported from Australia is to countries with no or low carbon taxes, such as Japan, Taiwan, China and Korea, it is a cost the climate will have to bear. For comparison, Australia's educational 'exports' from overseas students coming to Australia and courses run overseas accounted for approximately $19 billion in 2009–10,[82] but without the climate liability that comes with coal. Neither are export earnings really the same as national income. The Reserve Bank has noted that 'since the mining industry in Australia is majority foreign-owned, most dividends and retained earnings do not add to national income'.[83] The 'trickle-down' effect from corporate tax is also overstated, as in 2008–09 when the average rate of tax was just 13.9 per cent of the mining industry's gross operating surplus (a broad measure of profit used by the Australian Bureau of Statistics).[84] This is partly because of the generous tax concessions given to mining for capital spending.

While the coal industry boasts about the coal royalties it pays, in NSW and Queensland they represent approximately 3 per cent and

5 per cent respectively of state revenue.[85] In Victoria, coal royalties are so insignificant they are not even specifically mentioned in the budget papers. As the royalty rates in Queensland and NSW are based on the sale price, the recent drop in export value has seen state treasurers slash their forward revenue estimates. In Queensland, the government suffered a drop in royalties of almost $500 million in a single year due to the waning fortunes of the coal industry.[86] In NSW, lower thermal coal prices forced the Treasurer to cut the state's coal royalties estimate by just under $500 million in a year and $1.9 billion over a four-year forecast.[87] Just as with the job losses in the industry, the dramatic changes went unnoticed and were soberly reported in the backblocks of the budget papers by the conservative governments with the same resignation to market forces.

While the loss of royalty income from an export phase-out would be felt, the Queensland and NSW governments have already demonstrated that they can adapt, even though any falls from a coal exports phase-out would be less dramatic in a single year. Just as declining revenue from rapidly reducing smoking rates is no reason not to protect public health, coal royalties should be seen for what they are: a temporary windfall that masks the long-term costs they impose. Indeed, the costs of one major natural disaster exacerbated as a result of global warming can easily wipe out in one hit several years' worth of royalties. Cyclone Yasi in 2011 is estimated to have cost the public purse approximately $7 billion, with a similar amount borne by companies or individuals.[88]

Discussion of phasing out the coal industry invariably seems to spur industry supporters on to make exaggerated claims. However, it is worth remembering the dire predictions from the carbon tax debate. Opposition Leader Tony Abbott predicted that the tax would be 'absolutely catastrophic', 'wipe out jobs big-time, especially in the coal industry', create 'ghost towns' and 'discourage' investment in mining.[89] Yet just two months after the tax came into effect, he admitted that, 'the initial impact of the carbon tax may not be absolutely catastrophic.'[90]

Stephen Koukoulas, a former Senior Economic Adviser to Julia Gillard, noted that in the first eight months after the carbon tax came into effect real gross domestic product rose by over 1 per cent, full-time employment rose by over 50 000, the All Ordinaries Index on the ASX rose by 25 per cent and house prices rose by just under 3 per cent.[91]

So while a coal export phase-out would involve challenges for NSW and Queensland, they are survivable. It would mean the two states would need to gradually adjust their budgets, in the same way they have already had to do when faced with falling coal prices. Giving up spending public funds on CCS, for example, would save the state and federal governments billions of dollars. A gradual export phase-out would give employees and local communities a better chance of a smooth transition than relying on crude market forces. Export income from coal would fall over time, depending on how the global market responds, but a gradual depreciation of the Australian dollar would increase the value of foreign earnings from other sectors of the economy, including non-coal mining. Agriculture, manufacturing exporters, tourism and education sectors, to name a few – all significant regional employers in their own right – would breathe more easily. The economy would still expand in much the same way that it has after the introduction of the carbon tax and, on Treasury modelling, would still reach a projected doubling in size, just a year or two later than would otherwise have been the case.[92]

At a broader strategic level, a domestic economy that is highly energy efficient and based on renewables would give Australia smarter economic positioning in the world. With China teetering on the brink of making a historic shift away from coal, Australia is likely to be far better off as a supplier of skills and technologies needed in a carbon-constrained world than as an increasingly desperate hawker of coal.

Last drinks?

While Australia could go coal-free and be better off in the long run, none of the main political parties are jumping up to volunteer to lead the way. None of the peak business groups are going to either. Making the shift to a coal-free Australia will have to come from citizens making it clear that Big Coal's time is up.

In the last few years a new momentum has emerged in standing up to the coal industry, which transcends glib political categorisation. Radio announcer Alan Jones is someone who disproves the notion that opponents of coal projects have to be ardent believers in the threat of global climate change. He is both a strident climate change sceptic and an opponent of coal seam gas developments and some of the proposed coal mines in prime agricultural land. In WA, the ardently pro-mining conservative government led by Colin Barnett refused to consider the development of a coal mine near the wine, agricultural and tourism centre of Margaret River. After strong opposition from local residents, the Minister for Mines and Energy, Norman Moore, terminated all coal exploration licences within a 230-square kilometre zone around the town.[93] In Queensland, one of the first acts of the incoming government of Campbell Newman was to reject two proposed coal mines: the Felton Downs mine and the expansion of the Acland mine.[94] After community opposition to a proposed coal exploration project near Deans Marsh, west of Melbourne, Mantle Mining gave up and withdrew.[95] Faced with increasing costs and no banks willing to finance it, HRL gave up on its proposed new coal-fired power station in the Latrobe Valley when a federal grant was withdrawn.[96] Across large parts of northern NSW and southern Queensland, landholders alarmed at proposals for coal and coal seam gas have blocked access to farm properties with support from Lock the Gate group. To the horror of the gas and coal industry, in early 2013 the conservative NSW government imposed a ban on coal seam gas mining within 2 kilometre-wide zones around residential areas and designated industry clusters.[97] In the courts, several recent legal challenges have gone against coal companies. In one case, a small group of residents from the village of Bulga in NSW's Hunter Valley challenged and defeated a 200 million tonne coal project proposed by a subsidiary of Rio Tinto, one of the world's biggest coal miners.[98]

The gloom over the international coal market is claiming other victims. As international coal demand weakened, a proposed massive new coal export loader in Newcastle – dubbed T4 – has been cut back from 120 million times a year, first to 70 million tonnes and then to 25 million tonnes. Combined with a strong local community campaign against the new terminal, the proposal has been delayed and may be abandoned altogether.[99] Proposed coal export terminals and mines in Queensland have been delayed or shelved entirely. In the Kimberley, coal projects promoted by minor companies have largely withered away as coal prices have slumped.

These wins could be dwarfed by a tidal wave of mega-mines and infrastructure currently being enthusiastically backed by state and federal government, but even so, key supporters of the coal industry are nervous. Speaking at a mining industry conference in February 2011, former NSW Labor Treasurer and right-wing powerbroker, Michael Costa, recounted how 'at least two times in my period of government, there were serious proposals, almost at Cabinet level, to rule out any new coal mines in this state – That's how frightening the situation had got.'[100] What little public opinion polling on coal in Australia has been made public reveals the industry's fragile support base. One 2008 poll indicated just 18 per cent of respondents supported an increase in coal exports, one-third supported a decrease and just under half thought they should be at current levels.[101] In 2011, a poll for Network Ten found that 37 per cent agreed with the proposition that 'Australia's coal industry should be phased out by 2050', while 39 per cent disagreed and the remainder were unsure.[102]

Confronted with the alarming new climate projections and knowledge that the transition can be far less dramatic than the coal lobby would have us believe, it is time Australia stopped backing tired old Big Coal. Coal is killing a habitable planet and many of those who live near the power stations where it is burned. Australia could be a global leader in ending the reign of Big Coal, just as it is seen by governments around the world as a leader in curbing tobacco consumption. We didn't wait for an international agreement to introduce plain paper packaging for cigarettes; we acted unilaterally. Now governments around the world are following Australia's lead. Australia's journey from defender of tobacco to tobacco control champion was a long one, but its starting point was with a small group of public health champions who persisted in the face of what seemed to be the insurmountable strength of Big Tobacco. We don't have the luxury of decades to end the reign of Big Coal, but much of its power rests on the illusion that we have no choice. We do.

In international diplomacy, Australian Foreign Affairs commentators often claim that Australia 'punches above its weight' and is a well-respected middle power. But our greatest challenge lies ahead of us with global warming. Big Coal may, as Mike Bloomberg put it, be a 'dead man walking' in the United States, but it will require a huge groundswell of support to make it happen in Australia and beyond.

NOTES

INTRODUCTION

1. Paola Cassoni's stand is described in a number of newspaper stories: J. McCarthy, 'Mine mark "irreversible"', *Courier Mail*, 27 September 2011; B. Williams, 'Nature refuges get no protection from mining', *Courier Mail*, 21 October 2011; Australian Associated Press, 'Nature reserve owner "duped" on mine plan', 27 October 2011.
2. M. Wordsworth, 'Anger as Palmer mine threatens nature reserve', ABC News, 20 October 2011.
3. Australian Associated Press, 'Palmer confident of Qld mine approval', 1 December 2011.
4. S. Green, 'Nature no refuge against mining', *Central Queensland News*, 23 November 2011.
5. D. Robertson, 'Miners rally against resources tax', 'Lateline', ABC TV, 9 June 2010.
6. M. Priest, 'Hugs and duds, but no sign of thugs', *Australian Financial Review*, 10 June 2010.
7. D. Robertson, 'Miners rally', 'Lateline'.
8. D. Robertson, 'Miners rally', 'Lateline'.
9. B. Cubby, 'Coral wonderland at tipping point', *Sydney Morning Herald*, 14 July 2012.
10. B. Cubby, 'Records will keep tumbling with blistering heatwaves here to stay', *Sydney Morning Herald*, 9 January 2013.
11. AAP, 'Country getting hotter: Climate Commission', *Courier Mail*, 12 January 2013.
12. R. Garnaut, *The Garnaut Climate Change Review*, Cambridge University Press, Melbourne, 2008, p. 158.
13. E. Connolly & D. Orsmond, *The Mining Industry: From Bust to Boom*, Reserve Bank of Australia, 2011. On foreign ownership, see also Naomi Edwards, 'Foreign ownership of Australian mining profits', Briefing paper prepared for the Australian Greens.
14. ABS Labour Force, Australia, Detailed, Quarterly, Table 06, 'Employed Persons by Industry', February 2013. (Cat. No. 6291.0.55.003).
15. J. McCarthy, 'Foreign jobs lure for Alpha,' *Courier Mail*, 30 May 2012; M. Ludlow, 'Adani denies visa breaches', *Australian Financial Review*, 17 July 2012; 'Clive Palmer says he plans to seek an Enterprise Migrant Agreement', Federal Government Broadcast Alerts, 27 June 2012.
16. International Energy Agency figures in: Climate Commission, 'The Critical Decade: International Action on Climate Change,' p.18.
17. BP Statistical Review of World Energy 2012; http://www.bp.com/extendedsectiongenericarticle.do?categoryId=9041233&contentId=7075263
18. With every tonne of coal burnt producing 2.7 tonnes of carbon dioxide, our 581 million tonnes of exports will mean we export 1.56 billion tonnes of CO_2 each year. Saudi net oil exports today are currently around 8.7 million barrels a day (*CIA World Factbook*). This amounts to 3.175 billion barrels a year, which equates to around 1.36 billion tonnes of CO_2 produced annually, based on the conversion factor: 430 kilograms of CO_2 per barrel of crude oil (Source: US EPA, http://www.epa.gov/cleanenergy/energy-resources/refs.html). Thanks to Guy Pearse who first made this comparison in 'Australia's precious place in the coal industry's world', (http://www.guypearse.com/wp-content/uploads/2012/09/Pearse-Climate-Camp-Speech-Final.pdf.)
19. J. Gillis, 'Satellites show sea ice in Arctic is at a record low', *New York Times*, 28 August 2012.
20. J. Gillis, 'Satellites show sea ice', *New York Times*, 28 August 2012. A. Lowrey, 'Exports issue a warning as food prices shoot up', *New York Times*, 5 September 2012.
21. P. Canadell, 'The widening gap between present emissions and the 2 degree target',

The Conversation, 3 December 2012.
22 C. Le Quere et al., 'Trends in the sources and sinks of carbon dioxide', Nature Geoscience, 2, 831–36 (2009) cited in Climate Commission, The Critical Decade: Climate Science, Risks and Responses, Canberra, 2012.
23 J. Hansen, 'Coal-fired power stations are death factories', The Observer, 15 February 2009.
24 G. Pearse, High and Dry: John Howard, climate change and the selling of Australia's future, Viking/Penguin, Camberwell, 2007.
25 One-quarter (25 per cent) of steel production is produced from recycled steel by electric arc furnaces (secondary steel) and 6 per cent of primary steel production uses electricity. 'Fact Sheet – Energy', World Steel Association. (http://www.worldsteel.org/publications/fact-sheets.html)

1 THE REAL PRICE OF COAL

1 G. Orwell, 'Down the mine', in G. Bott (ed.) George Orwell: Selected Writings, Heinemann Educational, London, 1966, p. 34.
2 E. Duyker, Citizen Labillardiere, The Miegunyah Press, Carlton, 2004.
3 R. Gollan, The Coalminers of New South Wales: A History of the Union, 1860–1960, Melbourne University Press, Parkville, 1963, p. 5.
4 This early account relies on M. Diamond, 'Coal in Australian history' in P. Knights & M. Hood (eds), Coal and the Commonwealth: the Greatness of an Australian Resource, University of Queensland, 2009; E. Ross, A History of the Miners' Federation of Australia, Australasian Coal and Shale Employees Association, Sydney, 1970.
5 Evidence of William Evans, Assistant Colonial Surgeon, in J. W. Turner, Newcastle as a Convict Settlement, Council of the City of Newcastle, 1973.
6 Section 1301.0, Year Book of Australia 1982, Australian Bureau of Statistics, 1982.
7 Diamond, 'Coal in Australian history'.
8 G. Wilkenfeld, The Electrification of the Sydney Energy System 1881–1986, PhD Thesis, Centre for Environmental and Urban Studies, Macquarie University, 1989, pp. 62, 69.
9 Cited in G. Wilkenfeld, The Electrification of the Sydney Energy System 1881–1986, PhD Thesis, Centre for Environmental and Urban Studies, Macquarie University, 1989, p. 80.

10 Wilkenfeld, The Electrification of the Sydney Energy System 1881–1986, PhD Thesis, Centre for Environmental and Urban Studies, Macquarie University, 1989, p. 56.
11 Cited in J. Gunn, Along Parallel Lines: A History of the Railways in New South Wales, Melbourne University Press, Carlton, 1989, p. 136.
12 R. Broomham, First Light: 150 Years of Gas, Hale & Iremonger, Sydney, 1987, p. 9.
13 The Sun, 25 August 1911, cited in R. Broomham, p. 101.
14 This description based on J. D. Keating, The Lambent Flame, Melbourne University Press, Carlton, 1974, pp. 44–68.
15 Broomham, First Light, p. 86.
16 'Information for the assessment of former gasworks sites', Department of Environment and Conservation, Sydney South, 2005.
17 J. Hammond, 'High levels of toxins in Claisebrook fish', The West Australian, 4 June 2011.
18 G. Strong, 'Transforming an old site for sore eyes, skin, ears ...' Sunday Age, 19 October 1997.
19 I. Munro, 'Why isn't this man smiling?' The Age, 26 February, 2000.
20 M. Hele, 'Gasworks gets the all-clear', Courier Mail, 11 April, 2008.
21 ABC News, 'Public shouldn't pay for toxic clean-up', 24 July 2009.
22 'The Bulli Colliery explosion', Sydney Morning Herald, 25 March 1887.
23 Gollan, The Coalminers of NSW, p. 96.
24 Gollan, The Coalminers of NSW, p. 102.
25 Gollan, The Coalminers of NSW, p. 1.
26 Cited in E. Ross, A History of the Miners' Federation of Australia, p. 2.
27 Gollan, The Coalminers of NSW, p. 2.
28 P. Kinnear, 'The politics of coal dust: industrial campaigns for the regulation of Dust Disease in Australian coal mining, 1939–1949', Labour History, no. 80, May 2001.
29 Report of the Royal Commission upon the Safety and Health of Workers in Coal Mines, Government Printer, Sydney 1939, p. 297.
30 Report of the Royal Commission, p. 210.
31 Cited in Kinnear, 'The politics of coal dust', p. 72.
32 Kinnear, 'The politics of coal dust', p. 74.
33 W. M. Castleden, D. Shearman, G. Crisp & P. Finch, 'The mining and burning of coal: effects on health and the environment', Medical Journal of Australia, vol. 195, no. 6, p. 333.

34 G. Wilkenfeld and P. Spearitt, *Electrifying Sydney: one hundred years of EnergyAustralia*, EnergyAustralia, Sydney, 2004, p. 2.
35 Wilkenfeld & Spearritt, *Electrifying Sydney*, p. 4.
36 Wilkenfeld & Spearritt, *Electrifying Sydney*, pp. 6–9.
37 C. Doran, *Partner in Progress: A History of the Electricity Supply in North Queensland from 1897 to 1987*, James Cook University, Townsville, 1990, p. 5.
38 A. H. Lockwood, K. Welker-Hood, M. Rauch, & B. Gottlieb, *Coal's Assault on Human Health: A Report from Physicians for Social Responsibility*, 2009, p. vi; Physicians for Social Responsibility, founded in 1961, shared the 1985 Nobel Peace prize with International Physicians for the Prevention of Nuclear War for building public pressure to end the arms race. Since 1991 it has addressed issues of health and the environment, including global warming.
39 Lockwood et al., *Coal's Assault*, pp. 1–2.
40 Lockwood et al., *Coal's Assault*, p. 2.
41 Lockwood et al., *Coal's Assault*, pp. vii, 13.
42 Lockwood et al., *Coal's Assault*, pp. viii, 21ff.
43 Lockwood et al., *Coal's Assault*, pp. viii–ix.
44 A copy of the report can be obtained from: http://beyondzeroemissions.org/blog/coal-health-report-121023
45 N. O'Malley, 'Black marks on the health chart', *Sydney Morning Herald*, 19 March 2010.
46 Transcript, 'A dirty business', 'Four Corners', ABC TV, 12 April 2010.
47 Transcript, 'A dirty business' 'Four Corners'.
48 P. Cleary, 'Living in the dusty shadow of coal mining,' *Weekend Australian*, 28 January 2012.
49 T. Voight et al., 'Air pollution in Latrobe Valley and its impact on respiratory morbidity', *Australian and New Zealand Journal of Public Health*, vol. 22, no. 5, 1998.
50 W. M. Castleden, D. Shearman, G. Crisp & P. Finch, 'The mining and burning of coal: effects on health and the environment', *Medical Journal of Australia*, vol. 195, no. 6, 19 September 2011, pp. 333–5.
51 Drawn from export estimates in Bureau of Resources and Energy Economics, *Resources and Energy Quarterly*, September 2012, for thermal coal (165 mt) and metallurgical coal (151 mt).
52 'Australian bulk commodity exports and infrastructure – outlook to 2025', Bureau of Resources and Energy Economics, July 2012,

p. viii.
53 Cited in P. Cleary, *Too Much Luck: The Mining Boom and Australia's Future*, Black Inc, Collingwood, 2011, p. 9.
54 B. Lucarelli, *Australia's Black Coal Industry: Past Achievements and Future Challenges*, Program on Energy and Sustainable Development, Stanford University, 2011, p. 37.
55 Lucarelli, *Australia's Black Coal Industry*, pp. 59, 70.
56 Lucarelli, *Australia's Black Coal Industry*, pp. 70–7.
57 E. Connolly & D. Orsmond, *The Mining Industry: From Bust to Boom*, Reserve Bank of Australia, 2011, p. 17.
58 R. New et al., *Minerals and Energy, Major Development projects*, April 2011 listing, Australian Bureau of Agricultural and Resource Economics and Sciences, 2011.
59 New et al., *Minerals and Energy*, pp 3–4.
60 P. Cleary, 'Reforms must run faster', *The Australian*, 11 February 2012.
61 R. Garnaut, *The Garnaut Climate Change Review: Final Report*, Cambridge University Press, Port Melbourne, 2008, pp. 106, 113.
62 CSIRO & Bureau of Meteorology, *Climate Change in Australia: Technical Report 2007*, CSIRO, Melbourne; Lynch et al., *Defining the Impacts of Climate Change on Extreme Events*, 2008. (Cited in Garnaut 2008, p. 112.)
63 CSIRO, 'Projections of days over 35 degrees to 2100 for all capital cities under a no-mitigation case', Data prepared for Garnaut Review. See *Garnaut Review*, p.117.
64 Garnaut, *The Garnaut Climate Change Review*, p. 108.
65 W. Cai & T. Cowan, 'SAM and regional rainfall in IPCC AR4 models: can anthropogenic forcing account for southwest Western Australia winter rainfall reduction?' 2006. Cited in Garnaut, p. 109.
66 S. Lewis & S. Perkins, 'Hot spell is a taste of the future', *Courier Mail*, 30 March 2013.
67 Garnaut, *The Garnaut Climate Change Review*, p. 125.
68 Various examples drawn from Garnaut, *The Garnaut Climate Change Review*, Section 6, 'Climate change impacts on Australia'.

2 THE STRIP-MINING OF AUSTRALIA

1 ACSEA, *Australia Undermined: Coal in crisis*, Australian Coal and Shale Employees Association, Sydney, 1972, p. 42.

2   Hyder Consulting, 'Lower Hunter Transport Needs Study, Technical Paper 2, Freight Transport, 2008.
3   G. Ray, 'A coal change coming', *Newcastle Herald*, 12 February 2005.
4   This section is based on Debra Jopson, 'The Hunter: peephole to future of others', *Sydney Morning Herald*, 30 May 2009; G. Ray, 'Modern classic undermined by cruel fate', *Newcastle Herald*, 11 December 2009; Interview, Wendy Bowman, 13 July 2011.
5   L. Rickarby, 'Coal versus Hunter native title claims', *Newcastle Herald*, 25 June 2011.
6   D. Jopson, 'Colonial heritage undermined by big coal interests,' *Sydney Morning Herald*, 21 December, 2010.
7   A. Edwards, 'Historic property to stay on register', *Newcastle Herald*, 8 July 2011.
8   Hunter Thoroughbred Breeders Association, 'Submission to the NSW government's draft strategic regional land use plan', May 2012.
9   J. McCarthy, 'Second thoughts', *Newcastle Herald*, 19 May 2012.
10 Hunter Valley Wine Industry Association, 'Protecting the Hunter Valley from CSG Mining', March 2012, pamphlet, p. 5.
11 McCarthy, 'Second thoughts'.
12 Interview, Ian Napier, 10 August 2012.
13 A. Jaffe, 'Burdens of extraction: the growing coal mining industry in Australia's Hunter Valley', 17 May 2012, *Circle of Blue* online; Interviews, Martin Rush, 10 August, 6 September 2012.
14 G. Pearse, 'King Coal', *The Monthly*, May 2010.
15 ICAC, transcript of proceedings, 12 November 2012, http://www.icac.nsw.gov.au/investigations/current-investigations/categorylist/12/192. Accessed 10 December 2012.
16 K. McClymont, 'McGuigan negotiated with Moses Obeid', *Sydney Morning Herald*, 30 January 2013.
17 P. Cleary, 'Fertile areas face open cut threat', *The Australian*, 9 April 2011.
18 B. Robins & B. Cubby, 'Court ruling blocking mine access sidestepped', *Sydney Morning Herald*, 21 April 2010.
19 These figures were cited in a presentation by David Stolz from the Queensland Department of Infrastructure and Planning at the June 2009 Australia–Japan Coal Workshop.
20 Department of Employment, Economic Development and Innovation, 'Queensland's coal mines and advanced projects', June 2010.
21 D Stolz, June 2009 Australian–Japan Coal Workshop.
22 D. Passmore, 'Era of "black boot" brigade', *Sunday Mail*, 6 March 2011.
23 R. Barrett, 'Coal trial lashed for safety oversight' *The Australian*, 27 September 2012.
24 A. Klan, 'Approval given for ghost coal project', *The Australian*, 27 August 2011.
25 T. Fitzgerald, *Report of a Commission of Inquiry into Possible Illegal Activities and Associated Police Misconduct* (Fitzgerald Report), Queensland Government, Brisbane, 1989, pp. 102–04, 106–09.
26 P. Osborne, 'Nuttall seals own fate, puts Labor at risk', Australian Associated Press, 15 July 2009.
27 T. Rowling, 'There was anger in the air' *Queensland Country Life*, 17 March 2011.
28 M. Raats, 'Council backs mining protest', *The Chronicle*, 21 February 2011.
29 R. Barrett, 'Miners lose as frontline hits front lawn', *The Australian*, 23 June 2011.
30 M. Nadin, 'Town fired up over the fallout from coal', *The Australian*, 2 July 2011.
31 N. Onishi, 'Moving out one by one as Australia pursues coal', *New York Times*, 29 June 2010.
32 D. Houghton, 'Last man standing among ghosts', *Courier Mail*, 3 April 2010.
33 Houghton, 'Last man standing'.
34 Australian Society of Soil Scientists, Submission to the Senate Committee Inquiry into the Impact of Mining, September 2009.
35 J. McCarthy, 'Heritage lost in the spoils for Wandoan', *Courier Mail*, 14 July 2012.
36 C. Flately, 'Graziers, greenies take on Xstrata mine, Australian Associated Press, 22 August 2011.
37 M. McKenna, J. Owens, 'Xstrata a battle too far for grazier', *The Australian*, 22 August 2011.
38 House of Representatives Committee on Regional Australia, *Cancer of the Bush or salvation for our cities?*: Fly-in, fly out and drive-in, drive out workforce practices, Canberra, February 2013, Foreword, Section 5.
39 K. Carrington, R. Hogg & A McIntosh, 'The resource boom's underbelly: criminological impacts of mining development,' *ANZ Journal of Criminology*, vol. 44, no. 3, 2011
40 Australian Manufacturing Workers Union, Submission to the House of Representatives Standing

Committee on *Regional Australia Inquiry into Fly-In, Fly-Out Workers*, 2011, p. 17.

41 National Tourism Alliance, *Submission to the House of Representatives Inquiry into the use of 'fly-in, fly out' workforce practices in regional Australia*, 2011.

42 *Submission from the Moranbah Traders Association to the House of Representatives Inquiry into the use of 'fly-in, fly out' workforce*.

43 Carrington et al., 'The resource boom's underbelly', p. 342.

44 Greenpeace Australia Pacific, *Cooking the Climate, Wrecking the Reef; The global impact of coal exports from Australia's Galilee Basin*, Ultimo, Sydney, September 2012.

45 J. McCarthy, 'Foreign jobs lure for Alpha,' *Courier Mail*, 30 May 2012; 'Clive Palmer says he plans to seek an Enterprise Migrant Agreement', Federal Government Broadcast Alerts, 27 June 2012.

46 This section derived from M. Grudnoff, *Job Creator or Job Destroyer? An Analysis of the Mining Boom in Queensland*, Research paper, The Australia Institute, March 2012.

47 Cited in Grudnoff, *Job Creator or job Destroyer?*, p. 6.

48 Grudnoff, *Job Creator or Job Destroyer?* pp. 6–7.

49 Grudnoff, *Job Creator or Job Destroyer?*, p. 8 (citing research by the Queensland Government Office of Economic and Statistical Research).

50 B. Williams, 'Expert sinks harbour report', *Courier Mail*, 3 November 2011.

51 G. Lloyd, 'Dumping our heritage', *The Australian*, 6 March 2012.

52 These are conservative figures, drawn from PGM Environment, *Great Barrier Reef Shipping: Review of Environmental Implications*, 2012, Polglaze Griffin Miller and Associates, Safety Bay, WA, p. 68, part of the Abbot Point Cumulative Impact Assessment, commissioned by the port developers including Adani, BHP and GVK Hancock.

53 B. Williams, 'Reef at risk: full United Nations report attacks protection failures', *Courier Mail*, 21 June 2012.

54 PGM Environment, *Great Barrier Reef Shipping: Review of Environmental Implications*.

55 'Around the traps' (column) *Courier Mail*, 9 June 2011.

56 R. Millar, 'Brumby's dirty secret', *The Age*, 14 October 2009.

57 A. Carey, 'Not in our backyard, locals tell upbeat mining company' *The Age*, 18 October 2011.

58 T. Arup, 'Brown coal exports not yet viable: energy giant', *The Age*, 21 May 2012.

59 L. Taylor, 'Coalition U-turn on coal power station closures', *Sydney Morning Herald*, 21 July 2011.

60 N. Dalton, 'Coking coal mine first for Aborigines', *Cairns Post*, 28 April 2012.

3 THE BARONS' BOOM

1 E. Connolly, 'A ship where dreams come true': "Professor" Palmer launches Titanic II plans', *Sydney Morning Herald*, 27 February 2013.

2 R. Macklin & P. Thompson, *The Big Fella: The Rise and Rise of BHP Billiton*, Random House, p. 77; G. Pearse, 'Quarry vision: Coal, climate change, and the end of the resources boom', *Quarterly Essay* 33, Black Inc, 2009, p. 6.

3 Rio Tinto's predecessors were CRA Limited (Conzinc Riotinto of Australia Limited) and RTZ Corporation (Rio Tinto-Zinc Corporation), and before that the Consolidated Zinc Corporation. For more detail see: http://www.riotinto.com/annualreport2010/additional_information/history.html

4 For example, see: 'BHP boss Marius Kloppers says the carbon tax will hit coal investment', *The Australian*, 25 August 2011, online edition.

5 'Twenty Largest Registered Shareholders'. 2010 Annual Report – Rio Tinto, online edition.

6 'Australian banks financing coal and renewable energy', A research paper prepared for Greenpeace Australia, Profundo Research, Amsterdam, 2010, p. 6.

7 For example, see: K. Smith (PLATFORM), *Cashing in on Coal – RBS, UK Banks and the Global Coal Industry*, Report 2008, Bank Track, Friends of the Earth Scotland etc, p. 14; 'Australian banks financing coal and renewable energy', A research paper prepared for Greenpeace Australia, Profundo Research, Amsterdam, 2010, p. 6.

8 For example, see: 'Major shareholders', Investors Section, Xstrata.com. Accessed 11 April 2013 at: http://www.xstrata.com/investors/shareholder-centre/major-shareholders/

9 L. Armistead & J. Quinn, 'Sovereign wealth

funds back BlackRock move to acquire Barclays Global Investors', *The Telegraph* (UK), 12 June 2009; Z. Fattah, "'Kuwait fund made 40% return on Blackrock', Chief says', *Bloomberg*, 31 January 2010.
10 Glasenberg owns 15 per cent share of Glencore, which has 33.65 per cent of Xstrata's issued capital. The value of Glasenberg's assets only became clear once Glencore listed in 2011. More recently, Glencore and Xstrata agreed to merge. For more detail see 'Major Shareholders', Xstrata in Action, Xstrata, at: http://www.xstrata.com/investors/shareholder-centre/major-shareholders/
11 See J. Barber et al., *Resources and Energy Major Projects*, Australian Bureau of Resources and Energy Economics, Australian Government, Canberra, October 2012.
12 As of October 2012, the new coal projects in which these companies are involved handle an additional capacity of 187.6 million tonnes. Some projects involve other companies as partners. For more details see: J. Barber (et al.), *Resources and Energy Major Projects*, 2012.
13 In 2011–12 Australia exported around 300 million tonnes of coal. See: G. Armitage, *Resources and Energy Statistics – Annual 2012*, Australian Bureau of Resources and Energy Economics, Australian Government, Canberra, 2012, p. 46.
14 An additional 186 million tonnes of coal equates to an extra 502 million tonnes of CO_2. This compares with around 199 million tonnes of CO_2 emitted from coal combustion within Australia, mostly from coal-fired power stations. For more detail see: *CO_2 Emissions from Fuel Combustion – Highlights: 2012 Edition*, International Energy Agency, 2013, p. 51; G. Armitage, *Resources and Energy Statistics – Annual 2012*, 2012, p. 46.
15 'Best year yet for those at the top', BRW Rich 200, May–June 2011, pp. 32–3.
16 'The world's billionaires', *Forbes*, 2013, online.
17 Forbes Billionaires List; B. Woodward, 'Rinehart world's richest woman', *BRW* online, 4 June 2012
18 M. Wilkinson & J. Cohen, 'Gina Rinehart – the power of one', 'ABC Four Corners' (Program Transcript), 20 June 2012; J. Thomson, 'The five trials of Gina Rinehart', *The Power Index*, 17 July 2012.

19 Australia burns around 140 million tonnes of coal in total each year, about 75 million tonnes of which is black coal, the remainder being brown coal. See G. Armitage, *Resources and Energy Statistics – Annual 2012*. 2012, p. 46.
20 'First coal from Galilee Basin', Media Statement by Hancock Coal, 26 June 2011.
21 T. Grant-Taylor, 'Mining magnate Gina Rinehart set to sell off her coal assets in Queensland's Galilee Basin', *Courier Mail*, 21 May 2011, online edition.
22 'Cooking the climate: wrecking the reef – the global impact of coal exports from Australia's Galilee Basin', Greenpeace Australia Pacific, September 2012, pp. 25–7.
23 A. Fulwood, 'Gina Rinehart Alpha/Galilee Speech', Anne Fulwood – Town Square (Blog), 27 June 2011.
24 G. Rinehart, 'A call for action on serious challenges – Australian Resources and Investment', Occasional Paper, The Institute of Public Affairs, July 2012
25 'Gina Rinehart $2 a day', YouTube (video posted by Kerry Seebohm)
26 G. Rinehart, 'Let's get back to our roots', *Australian Resources and Investment*, September 2012.
27 R. Lawson, 'Gina's poetic swipe at critics', *Perth Now*, 11 February 2012.
28 'Open Letter – Let us bring in workers to a Special Northern Zone', Australians for Northern Development & Economic Vision, 29 April 2010.
29 'BHP chairman confirms ANDEV warning on Australian mining', Media Release by the Institute of Public Affairs, 17 May 2012; C. Hamilton, 'Mining in a new vein', *The Age*, 2 February 2012.
30 G. Rinehart, *Northern Australia and then some: changes we need to make our country rich*, Self-published. 2012; Other sources include: Hancock Prospecting, and http://www.andev-project.org/ – accessed 11 January 2013.
31 J. Nova, 'Influential people are getting the message: Gina Rinehart explains the science of climate change', Joanna Nova (blog), 7 December 2011.
32 T. Arup, 'Academics campaign against Lord Monckton's Lang Hancock lecture', *The Age*, 30 June 2011; R. Manne, 'Lord Monckton and the Future of Australian Media', *The Monthly* (Blog), 8 February 2012; 'Rinehart appoints Plimer to another board', The Power

Index, 16 February 2012.
33 'Cooking the climate: wrecking the reef ...' Greenpeace, 2012, pp. 29–31.
34 The estimated combustion emissions associated with the coal produced by the Alpha North and China First projects alone would be over 171 million tonnes of CO_2 annually. This is more than 25 per cent of current global aviation emissions of around 670 million tonnes, see: 'Cooking the climate: wrecking the reef', Greenpeace, 2012; and 'Fact Sheet: Climate Change', International Air Transport Association, 2013.
35 Presentation by N. Harris, Waratah Coal, Coaltrans Australia Conference, Brisbane, August 2011, p. 7.
36 'Clive Palmer', *Gold Coast Bulletin*, 7 June 2008.
37 Clive Palmer – Extended Interview Transcript – Part One, 'Australian Story' (Program Transcript), ABC Television, 7 May 2012.
38 In 2010, for example, then Queensland Premier Anna Bligh and her Treasurer Andrew Fraser, settled a defamation action launched 18 months earlier by Palmer after they accused him of buying the Liberal–National Party with donations. See S. Wardill, 'Clive Palmer drops defamation action against Premier Anna Bligh', *Courier Mail*, 6 August 2010; R. Barrett, 'Clive Palmer, Anna Bligh make up', *The Australian*, 7 August 2010.
39 'Mr. Clive Frederick Palmer', CV for the Brisbane Mining Club, 5 September 2011.
40 CEDA 'In 2020 – A conversation with Clive Palmer (Meet the Speaker)', CEDA, 7 April 2011; 'China and the Australian Resources Boom presented by Clive Palmer' (About the Speaker), Australian Property Institute, August 2011 'Breakfast with Professor Clive Palmer', Griffith University, September 2010.
41 T. Allard & S. Washington, 'Tracing Palmer's wealth – a *Titanic* undertaking' *Sydney Morning Herald*, 12 May 2012. J. Walker, 'Clive Palmer: having it all', *The Australian*, 18 August 2012.
42 http://www.mineralogy.com.au/directors.html; http://www.waratahcoal.com/directors-and-management.htm https://twitter.com/search?q=clive+palmer
43 M. Atkin, 'Palmer blasts "poisonous" coal seam gas industry', ABC News Online, 29 August 2011.
44 'Q&A Climate Debate (Program Transcript)',

'Q&A', ABC Television, 26 April 2012.
45 E. Connolly, 'A ship where dreams come true': "Professor" Palmer launches *Titanic II* plans', *Sydney Morning Herald*, 27 February 2013.
46 'Tinkler down but still at the table', *BRW*, 27 September 2012, online edition.
47 S. Tasker, 'Tinkler punts on Rio coal project', *The Australian*, 5 November 2009, online edition.
48 G. Roberts, 'Tinkler wealth plunges more', *The West Australian*, 1 March 2013, online edition.
49 T. Reilly, 'The fortune and fury of a young tycoon', *Sydney Morning Herald*, 10 October 2010, online edition.
50 P. Manning, '$2.7 million debt: Tax Office moves to wind up Tinkler's teams', *Sydney Morning Herald*, 13 December 2012.
51 'Tinkler pulls out of Whitehaven takeover', AAP, 24 August 2012.
52 '2011 Northern Region Nominees – Peter Bond: Linc Energy', Entrepreneur of the Year Program, Ernst & Young, See: http://eoy.ey.com.au/peter-bond-linc-energy/w1/i1610576/
53 M. Smith, 'License to Drill: Peter Bond is on a mission', *BRW*, 22 November 2012, online edition.
54 A. Fraser, 'Linc Energy and Adani Group strike record deal', *The Australian*, 4 August 2010, online edition.
55 Teresa Coal Project – Initial Advice Statement, Report by GHD Consultants, Linc Energy, October 2011, p. 9.
56 'Linc Energy "Big Squeeze" TV commercial', YouTube (uploaded by Chris Adams), viewed 29 January 2012.
57 'Secret millionaire in Redfern Waterloo', YouTube (uploaded by redfernoralhistory), viewed 16 January 2010.
58 'The Diesel Dash', Rough Diamond Media, 17 November 2011.
59 G. Lower, 'Carbon tax 'just revenue raising', says Linc Energy chief Peter Bond', *The Australian*, 25 September 2010, online edition.
60 'Carmichael Royalty Update', ASX Announcement by Linc Energy, 6 December 2012.
61 T. Grant-Taylor, 'Xstrata's hope for coal in Wandoan', *Courier Mail*, 1 February 2010, online edition.
62 'QCoal Projects', Fact Sheet by QCoal,

available at: http://www.qcoal.com.au/files/qcoal_projects_poster.pdf

63 There are perhaps 4.5 million registered hybrids worldwide (late 2012), but this is up from 1 million in 2007 and 2 million in 2009; So if we assume hybrids get 5t CO_2 better per annum than regular alternative, that's 22.5 million t CO_2-e a year saved now, but about 10 million per annum back to 2009, about 5 million in 2007, and progressively much less before. The cumulative CO_2 saving total is likely to be less than 70 million tonnes in total – and roughly similar to the 62 million tonnes of CO_2 emanating from QCoal's new projects. US Department of Energy; See also: G. Pearse, 'Mine coal, sell coal, repeat until rich' Speech at the Woodford Folk Festival, December 2011, pp. 9, 17.

64 'Malaysia born billionaire in Australia – Sam Chong' *Kuala Lumpur Post*, 27 May 2012, online edition

65 K. McClymont, 'Magnificent Seven's money-making machine', *Sydney Morning Herald*, 8 December 2012, online edition.

66 J. Thomson, 'Rich Pickings: I'd like to thank the economy', *Business Spectator*, 14 December 2012, online edition.

67 K. McClymont, 'Magnificent Seven's money-making machine', *Sydney Morning Herald*, 8 December 2012, online edition.

68 L. Besser & K McClymont, 'Macdonald would get $4m from rigged tender, ICAC told', *Sydney Morning Herald*, 7 February 2013; J. Wells & P. Lloyd, 'ICAC hears Macdonald was offered millions in kickbacks' ABC News Online, 7 February 2013

69 P. Manning, 'The boy with the black stuff', *Sydney Morning Herald*, 12 May 2012, online edition.

70 P. Manning, 'Haggarty takes shot at carbon tax', *Sydney Morning Herald*, 12 May 2012, online edition.

71 Having been established in 1999, at the time Whitehaven listed in 2007, the share price was less than $2, compared with $2.39 today. Haggarty acquired most of his shares between April and October 2008 when the share price was between $2.80 and $4.50. During speculation over the Tinkler takeover in 2012, the share price went higher than $7; ASX; Whitehaven Coal; 'Load up the golden shovel', *BRW*, 23 March 2011, online edition.

'Wood Mackenzie's Second Coal Acquisition – Barlow Jonker – Brings Together Global Expertise in the Oil, Gas, Power & Coal Sectors' Media Release by Woods Mackenzie, 9 July 2007

72 http://www.woodmacresearch.com/cgi-bin/wmprod/portal/corp/corpPressDetail.jsp?oid=834107 http://www.asx.com.au/asxpdf/20120928/pdf/4291vm8n9j8rv7.pdf

73 'Bandanna Energy draws Chinese, Indian suitors: report', *Sydney Morning Herald*, 18 April 2011; 'Bandanna Energy bidding to be dominated by Asian buyers', *The Australian*, 14 February 2011, online edition.

74 Bandanna Energy; See also: 'Cooking the climate: wrecking the reef – the global impact of coal exports from Australia's Galilee Basin', Greenpeace Australia Pacific, September 2012, pp. 2, 3.

75 Australian Stock Exchange

76 15 million tonnes of coal equates to around 40 million tonnes of CO_2 from combustion emissions. This is about 9 million cars at an average of 4.5t CO_2 per car.

77 Biological Farming Systems Soil Carbon Tour (BFSSCT) Vic - 6th August 2009. Based on Tour Booklet from BFSSCT held 11–12 May 2009 [Revised Version 4th August 2009]' Report published by Sustainable Business Australia, 2009 p. 23. See also: http://www.sba.asn.au/sba/pdf/BioCCS-biological-farmers.pdf

78 A statement to on the Baker and McKenzie website quoted a senior partner as saying 'The acquisition of Cascade Coal by White Energy is an exciting step by White Energy in achieving its objective to develop conventional coal assets. We are very happy to be advising on a mutually beneficial transaction for both parties'. See: 'Baker & McKenzie Advises Cascade Coal Shareholders on Sale to White Energy', Deal Announcement, Baker & McKenzie, 15 February 2011.

79 SGH website at http://www.sevengroup.com.au/about-westrac/ Accessed 24 July 2012.

80 Seven Group Holdings, SGH *Annual Report 2012*, 2013, p. 6; 'WesTrac exceeds earnings expectations – Seven Group Holdings', Perth Now, 25 August 2011, online edition.

81 SGH, *Annual Report 2012*, 2013, p. 7.

82 SGH, *Annual Report 2012*, 2013 p. 10; Climate Killer Banks, 2011 p. 35.

83 A. O'Brien, 'Keep China close, says Kerry Stokes', *The Australian*, 13 July 2011, online edition.
84 A. White, 'Stokes not worried about China', *The Australian*, 31 August 2012, online edition; B. Packham & J. Massola, 'Julia Gillard comes face to face with business attacks on carbon, budget surplus', *The Australian*, 16 May 2011, online edition; C. Kerr, 'Do a better selling job on carbon plan, bosses tell PM', *The Australian*, 17 May 2011, p. 3, online edition.
85 In 2012, the mining-dominated Westrac Australia business generated $387.1 million in earnings before interest and tax (EBIT). SGH's investments in media generated less than a third of that at $116.1 million. See: SGH Annual Report 2012, 2013 pp. 7, 9; C. Kruger, 'Caterpillar powers Seven profit', *Illawarra Mercury*, 26 February 2013, online edition.
86 C. Williamson, 'Kings team up for Cup bid with Moatize', Wide World Of Sports – Channel Nine, 3 November 2008; S. Brem, 'First boozeday in November', This Racing Game (Blog), 2 November 2008
87 C. Baxter, 'Exchange place to be if Moatize gets up in Melbourne Cup', *Townsville Bulletin*, 1 November 2008.
88 'Corporate Presentation – Riversdale Mining Limited', Speech by Michael O'Keeffe, Sydney Mining Club, December 2008.
89 E. Robinson, 'Burbank Contrarian Bets on Saudis Show Why Oil Reality Helps Return 23.6%', *Bloomberg Markets Magazine*, 8 June 2011.
90 S. Tasker, 'Magnates attracted to a resources comeback', *The Australian*, 11 January 2012, online edition.
91 D. Winning, 'Riversdale steps up plans for listing', *The Australian*, 12 January 2013, online edition; See also: M. Chambers, 'Rio's $2.9 bn Riversdale writedown blamed on ambitious expansion and technology', *The Australian*, 31 January 2013, online edition.
92 D. Nayyar & U. Mahurkar, 'Billionaire who beat the system – A look into profile of billionaire Gautam Adani', *India Today*, 20 August 2011, online edition.
93 J. Derasari, 'India's Energy Needs and Adani Group in Australia', Presentation to Coaltrans Australia conference, by Jignesh Derasari, August 2011, pp. 9,13.
94 Nayyar & Mahurkar, 'Billionaire who beat the system', *India Today*, 20 August 2011.
95 According to the latest BREE report (2012) Australia had 29.7 gigawatts of installed coal fired power capacity in 2009-10; 22.3GW of black coal-fired power, and just over 7 gigawatts of brown coal-fired power. Energy In Australia – 2012, Australian Bureau of Resources and Energy Economics, Australian Government, Canberra, February 2012 p. 37.
96 See G. Pearse, 'Mine coal, sell coal, repeat until rich'. Speech by Guy Pearse at the Woodford Folk Festival, December 2011 pp. 3, 6.
97 'Adani Power Plant in Chindwara Illegal. Attack on Dr. Sunilam and Kisan Sanghrsh Samiti Condemned', Press Note – National Alliance of Peoples Movements, undated. See: 'Adani's record of environmental destruction and non-compliance with regulations', Research Briefing by Greenpeace Australia-Pacific, 11 February 2013.
98 'Adani defies trend with $10 bn Galilee project', *The Australian*, 3 December 2012, online edition.
99 Ian S. 'Fuel shortage hits GVK Power's expansion in Andhra', TwoCircles.net, 5 December 2012.
100 84 million tonnes of coal equates around 226 million tonnes CO_2. So, 11.3 times as much CO_2 is added by the Gillard-government-backed GVK mines in the Galilee Basin as their abandoned pledge to fund the closure of 2 gigawatts of Australia's dirtiest coal fired power might have saved.
101 P. Stafford, 'Gina Rinehart close to GVK deal, but sparks controversy by flying MPs to wedding of Indian industrialists', Smart Company, 17 June 2011.
102 A. Phillips, C. Arthur & J. Harris 'Galilee Basin test pit opens coal mining future', ABC News (Western Queensland), 9 November 2010; 'Opening of the Hancock Coal Test Pit', Media Release by Martin Ferguson, Minister for Resources and Energy, 6 November 2010.
103 'Alpha Coal mine and rail project approved', Media Release by Tony Burke, Minister for Sustainability, Environment, Water, Population and Communities, 23 August 2012; See also 'Approval Decision – Alpha Coal Mine and Rail Proposal, Galilee Basin, Queensland (EPBC 2008/4648).
104 S. Seth, GVK's $10bn coal proposal gets Australian government nod, Mineweb

Australia, 23 August 2012.
105 'China's Tycoons – Profiles of 100 top business leaders' (2nd edn), Week in China, HSBC, August 2012 p. 65.
106 According to BREE, some 60 million tonnes of black coal is used for power generation in Australia. This translates to around 160 million tonnes of CO_2 at a typical conversion rate of 2.7 tonnes of CO_2 per tonne of coal; G. Armitage, *Resources and Energy Statistics – Annual 2012*, 2012, p. 46.
107 Project China Stone – *Initial Advice Statement for Macmines Austasia Pty Ltd*, Hansen Bailey Environmental Consultants, 14 September 2012, p. 5; M. Chambers, 'Chinese push for $3bn-plus coalmine in Queensland', *The Australian*, 5 March 2013.
108 According to the latest figures from BREE, 22.34 gigawatt of black coal-fired power generated 124.5 TWh in 2009–10. In rough terms, this equates to 5.58 TWh per GW of installed black coal-fired capacity. An 800 MW (.8 GW) coal-fired power station might be expected to generate around 4.4 TWh. Meanwhile, in 20010–11 wind power generated 5.8 TWh and Solar PV generated 0.85 TWh for a combined total of 6.65 TWh. Thus, even allowing for additional growth in solar since, the proposed Meijin coal-fired power station would erase at least half the emissions being saved by wind and solar. See: *Energy In Australia*, 2012, pp. 34–7; 'Australian electricity generation, by fuel type – physical units – 2010/11', Australian Bureau of Resources and Energy Economics, Canberra, 2012.
109 China Environmental Energy Investment Limited, Memorandum of understanding in relation to a possible acquisiton, 2012; Current and Historical Company Extract, Macmines Austasia, Australian Securities and Investments Commission, 23 November 2012.
110 According to the *Annual Report of the Queensland Government Coordinator General*, 'Geoff Dickie—Appointed Deputy Coordinator-General (Infrastructure and Economic Development), 22 May 2008 for a period of three years and Deputy Coordinator-General (Project Assessment and Attraction), 30 August 2010 to 21 May 2011.' ASIC company documents show that he was appointed as a director of Macmines Austasia on 30 June 2011. Macmines Austasia, Current and Historical Company Extract, Australian Securities and Investments Commission, 23 November 2012. See also: (Appendix 1: Coordinator-General's report detail) DEEDI Annual Report 2010–11, Department of Employment, Economic Development and Innovation, Queensland Government, 2011 p. 129.

4 CHARM OFFENSIVE

1. 'Coal in the Commonwealth study: points to the future of coal in Australia', Media Release, University of Queensland, 28 October 2009.
2. 'This is our story – Ken Lamb – 60 second version', YouTube (video uploaded by Ausminingstory), viewed 17 April 2013.
3. Minerals Council of Australia, 'Australian Mining: This is our story' website, 'Agriculture and Mining's Story', http://www.thisisourstory.com.au. Accessed 17 April 2013.
4. Minerals Council of Australia, 'Australian Mining: This is our story' website, 'John's Story'; 'Bruce's Story'. Accessed 17 April 2013.
5. Minerals Council of Australia, 'Australian Mining: This is our story' website, 'Heather's Story'. Accessed 17 April 2013.
6. S. Marshall, 'Hearts, minds and hip pockets: how the resources industry aims to win over ordinary Australians', *The Conversation*, 11 October 2012.
7. Minerals Council of Australia, 'Australian Mining: This is our story', 'Paul's Story'; Accessed 17 April 2013.
8. Minerals Council of Australia, 'Australian Mining: This is our story', 'Daniel's Story'; Accessed 17 April 2013.
9. Lawrence Creative, 'This is our story – sizzle reel', http://www.lawrencecreative.com.au Accessed 16 April 2013
10. Regarding the film, *Red Dog*, Woodside, Rio Tinto and WesTrac provided everything from free accommodation to helicopters to film extras, with senior executives like Rio Tinto CEP acknowledging it was 'an exciting opportunity to showcase our industry'. See G. Readfearn, 'Hit movie *Red Dog* and its mining industry funding', DeSmog Blog, 9 September 2011; ScreenWest, 'Resources sector backs *Red Dog*, 10 March 2010, online. See also 'Coal Chain Gatecrash', Golden Target Award Entry by Repute Communications & Associates – Client: Port Waratah Coal

Services, University of Technology Sydney, 2009
11. B. Jabour, 'Clive takes a step to the left to deliver hope for sacked public servants', *Brisbane Times*, 24 August 2012, online edition.
12. Xstrata Coal website, 'What Matters – Xstrata Corporate Social Involvement Program 2012' http://www.xstratacoal.com. Accessed 13 April 2013; Rio Tinto website, 'Rio Tinto Coal Australia – Where does our Funding Go?'. Accessed 13 April 2013.
13. Rio Tinto Coal Australia, 'Where does our Funding Go?' 'The Edible Schoolyard: Applicant: St Catherine's Catholic College'.
14. Xstrata Coal Wandoan Project, 'Wandoan Community Fund', Xstrata.com.au, Accessed 14 April 2013.
15. Rio Tinto Coal Australia, 'Where does our Funding Go?' 'Arts (Hip Hop Dance) Workshops – Applicant: Contact Inc', 'Upper Hunter Show Band Tour: Applicant: Band Tour.
16. Rio Tinto Coal Australia, 'Where does our Funding Go?' 'Black Magic: Applicant: Toni Fisher, Magenta-Rae Fisher and Lacey Malone', 'Modelling School Scholarship: Applicant: Chloe Griffiths'.
17. Rio Tinto Coal Australia, 'Where does our Funding Go?' 'Aspiring Ballerina: Applicant: Cheyanne Dunstan'.
18. Rio Tinto Coal Australia, 'Where does our Funding Go?' 'Nebo swimming instructor for babies and toddlers: Applicant: Brigalow Babies Playgroup'.
19. Sponsored by Peabody. Minutes of the Wilpinjong Community Consultative Committee held at Wilpinjong Coal Peabody Boardroom, 7 November 2011 Upper Hunter Wine & Food Fair: 'Applicant: Denman Chamber of Commerce', Recreation, Rio Tinto
20. Xstrata, 'What Matters – Xstrata Corporate Social Involvement Program', 2012, p. 11.
21. Aurizon, 'Easter Clinic', Community Partners, Aurizon.com.au; Newcastle Knights website, 'QR National Train with the Knights', Newcastle Knights News, 6 July 2012.
22. Xstrata, 'Xstrata Corporate Social Involvement Program 2010', p. 31.
23. Rio Tinto Coal Australia, 'Where does our Funding Go?', 'NRL Sport Sponsorship – Applicant: Keenan Yorston'; See also: 'All Blacks carnival booming', Townsville Bulletin, 8 October 2011.
24. Xstrata, 'Wandoan Community Fund', Xstrata Coal Wandoan Project, Xstrata.com.au, Accessed 14 April 2013.
25. Xstrata, Corporate Social Involvement brochure, 2012, p. 4; See also 'What Matters – Corporate Social Involvement Program 2012', p. 20.
26. Xstrata, 'Corporate Social Involvement brochure, 2012; 'NSW Mining Industry contributes $270,000 to Westpac Rescue Helicopter Service', Media Release, NSW Minerals Council, 24 July 2012.
27. A. Hume, 'Mudgee Hospital secures new helipad path from Xstrata funding', *Mudgee Guardian*, 27 March 2012.
28. M. Levy, 'Clive Palmer's helicopter saves 60 from floods', *Brisbane Times*, 13 January 2011; Aurizon Community Giving, qrnational.com.au, Accessed January 2013
29. See also Xstrata, 'What Matters – Corporate Social Involvement Program 2012', p. 22.
30. Minutes of the Wilpinjong Community Consultative Committee held at Wilpinjong Coal Peabody Boardroom, 7 November 2011.
31. C. Walker, 'Rural Health Centre official', invitational opening, Namoi Valley Independent, 13 September 2012.
32. 'Xstrata Coal extends grant for AEIOU Regional Autism Program', Media Release, AEIOU Foundation for Children with Autism, 7 November 2011; 'Community Foundations – BHP Billiton Cannington Community Fund', 'Working Wonders for Sick Kids', workingwonders.com.au. Accessed 14 April 2013; 'Flying Doctor Queensland', Newsletter of the Royal Flying Doctor Service (Queensland Section), RFDS, November 2010, p. 5; 'Region to benefit from asthma research partnership', Media Release, University of Newcastle, 6 May 2008.
33. Rio Tinto Coal Australia, 'Where does our Funding Go?' 'Tourism Initiative Development Strategy – Applicant: Belyando Shire Council; $20,000.00', 'Tourism Development Officer – Applicant: Belyando Shire Council, $25,000.00', 'Tourism/Museum Development Officer – Applicant: Belyando Shire Council, $25,000.00'
34. Rio Tinto Coal Australia, 'Where does our Funding Go?' 'Feasibility Study – Applicant: Hunter Brands Steering Committee'.
35. BHP Billiton, 'BHP Billiton brings soothing Symphony to drought-stricken farmers', News Release, 11 July 2008.

36 Environment Services – Mining, Department of Environment and Heritage Protection, Fitzroy River Catchment 2011–12 Wet Season Report, Queensland Government, Brisbane, 2011; Rio Tinto Coal Australia, 'Where does our Funding Go? 'Turtle Talks – Applicant: Greening Australia'.

37 Minutes of the Wilpinjong Community Consultative Committee held at Wilpinjong Coal Peabody Boardroom, 7 November 2011.

38 'Mining, sew work, and STIs – why force a connection', *Australian Mining*, 9 August 2012; J. Scott & V Minnichiello, 'Mining, sex work and STIs: why force a connection?' *The Conversation*, 8 August 2012; 'Discrepancies to blame for STIs', *The Daily Mercury*, 28 May 2012 ; 'Sex workers deny FIFO STI stink', AAP, 25 May 2012.

39 For more on these concerns see the recent report stemming from a Parliamentary Inquiry into FIFO and DIDO practices and their impact on communities, *Cancer of the bush or salvation for our cities?, Report by the House of Representatives Standing Committee on Regional Australia*, Commonwealth of Australia, February 2013; for more background on the problems associated with FIFO see, for example, 'The dark side of the boom' Background Briefing, ABC Radio National, 25 March 2012 ; see also Casualties of the boom', 'Four Corners', ABC Television, 28 May 2012 ; and 'Remote boom and gloom', Background Briefing, ABC Radio National, 10 May 2009.

40 Rio Tinto Coal Australia, 'Where does our Funding Go?' 'Building better business in the Hunter – Applicant: Hunter Region Business Enterprise Centre'.

41 Rio Tinto Coal Australia, 'Where does our Funding Go?' 'Better Business Workshops – Applicant: Commerce Queensland'.

42 'Rio Tinto Coal Australia, 'Where does our Funding Go?' 'Clermont BioFuel and BioEnergy Feasibility Study – Applicant: Isaac Regional Council', 'Blue Metal Quarry Pre-Feasibility Study – Applicant: Belyando Economic Development Group', 'Traditional indigenous jewellery – Applicant: Marabisda Inc.'

43 Rio Tinto Coal Australia, 'Where does our Funding Go?' 'Succession: The future of the family business – Applicant: CHRRUP'

44 Rio Tinto Coal Australia, 'Case Study: Clermont Preferred Futures recognised at National Awards for Economic Development Excellence, Sustainable Development Report 2011, 2011.

45 Anglo American, 'Anglo American delivers new homes for Moranbah', *Moranbah Community News*, Issue 1 June 2012, p. 2 ; 'Casualties of the boom', 'Four Corners', ABC Television, 28 May 2012.

46 'The miners are coming to quiet Alpha', *CQ News*, 28/9/11

47 Rio Tinto Coal Australia, 'Where does our Funding Go?, 'Affordable Housing Strategy' – Applicant: Nebo Shire Council'.

48 'Drive in drive out culture poses safety questions', '7.30 Report', ABC Television, 23 May 2012.

49 Rio Tinto Coal Australia, 'Where does our Funding Go?, 'F3 Freeway – Applicant: Singleton Shire Councils'; 'Local Councils Supporting the F3' – Applicant: Hunter Economic Development Corporation'

50 Xstrata, 'What Matters – Xstrata Corporate Social Involvement Program 2012, 'Xstrata funds the John Hunter Trauma Hospital Unit', p. 21.

51 Rio Tinto, 'Rio Tinto in Mackay creates hope for cancer patients this Christmas', Media Release, 14 December 2009 .

52 See 'Other' category at Rio Tinto Coal Australia, 'Where does our Funding Go?'

53 'Funding opportunities for Alpha/Jericho District Clubs/Associations – GVK/Hancock Community Support Program: Notification of grants available for 2012/13 Barcaldine Council. http://www.barcaldinec.qld.gov.au/news-and-events/-/asset_publisher/v3RU/content/

54 Rio Tinto Coal Australia, 'Coal & Allied extends Council grants officer partnerships', Media Release, 10 November 2011.

55 See, for example, 'Westpac Rescue Helicopter Black Coal Challenge Cup', Media Release, CFMEU (Mining & Energy Division); and the CFMEU's major sponsorship of the Outback Rugby League: 'The Outback Rugby League sponsorship Coup', Sportingpulse.com Accessed 16 April 2013

56 FFA terminates Palmer's license', FFA News Story, 29 February 2012 Statement, 'Palmer v FFA - Where it all went wrong' *ABC News*, 29 February 2012; R. Gatt & J. Owens, 'Clive Palmer opens up new war with FFA', *The Australian*, 19 April 2012.

57 SMI funding partners in 2010–2011 included

Anglo American, Ashton Coal, BHP Billiton, Centennial Coal, Leighton, Macarthur Coal, New Hope Corporation, Peabody Energy, Rio Tinto, Stanmore Coal, Stanwell, Tata, Thiess, Vale, Wesfarmers, Whitehaven and Xstrata. See: *Biennial Report 2010–11*, Sustainable Minerals Institute – University of Queensland, Brisbane, 2011, pp. 52–3.

58 P. Knights & M. Hood, *Coal and the Commonwealth: The Greatness of an Australian Resource*, University of Queensland, Brisbane, 2009.

59 University of Queensland, 'Coal in the Commonwealth study: points to the future of coal in Australia', Media Release, 28 October 2009.

60 M. Diamond, 'Two centuries of coalminers' toil helped build the nation's wealth', *The Australian*, 16 July 2011, online edition.

61 'Coal and the Commonwealth', Golden Target Award Entry by Rowland – Client: Peabody Energy Australia, University of Technology Sydney, 2010.

62 University of Queensland, 'New UQ research centre to focus on coal seam gas', Media Release, 2 November 2011.

63 Centre for Coal Seam Gas, Vision: To become the World Leader in CSG Education and Research, Presentation by Chris Moran, Director Sustainable Minerals Institute, Interim Director, Centre for Coal Seam Gas, University of Queensland, n.d. pp. 22–5.

64 Centre for Coal Seam Gas, Vision: To become the World Leader in CSG Education and Research, Presentation by Chris Moran p. 23; Centre for Coal Seam Gas, 'Centre Governance', Sustainable Minerals Institute, University of Queensland; G. Readfearn, 'Expect scepticism over gas industry-funded research, *Brisbane Times*, 21 November 2012.

65 Centre for Coal Seam Gas, 'Frequently Asked Questions about CCSG', Sustainable Minerals Institute, University of Queensland.

66 University of NSW, 'Our Partners – Corporate', Australian Centre for Sustainable Mining Practices, http://www.casmp.unsw.edu.au. Accessed 16 April 2013.

67 University of NSW, 'Mitsubishi backs sustainable mining', Media Release, 29 May 2009.

68 Rio Tinto Coal Australia, 'Scholarships – Want to get paid to learn? Find out more about our scholarships', http://www.riotintocoalaustralia.com.au. Accessed 17 April 2013.

69 Xstrata, 'Our education partners', http://www.xstratacoal.com. Accessed 17 April 2013.

70 Rio Tinto Coal Australia 'Scholarships -- Want to get paid to learn? Find out more about our scholarships'.

71 Queensland Minerals and Energy Academy, 'Our Schools', Queensland Government ; See also, 'What is coal?', Fact Sheet, Oresome Resources – Minerals and Energy Education, Queensland Resources Council, Brisbane, http://www.oresomeresources.com. Accessed 17 April 2013.

72 Further detail search the database of ACARP-funded research: Australian Coal Association Research Program, 'ACARP Reports', http://www.acarp.com.au/reports.aspx. Accessed 17 April 2013.

73 D. Franks, D. Brereton, C. Moran, T. Sarker & T. Cohen, *Cumulative Impacts – A Good Practice Guide for the Australian Coal Mining Industry*, Centre for Social Responsibility in Mining & Centre for Water in the Minerals Industry, Sustainable Minerals Institute, University of Queensland. Australian Coal Association Research Program, Brisbane, 2010; Australian Coal Association Research Program, 'Cumulative Impacts: A Good Practice Guide For the Australian Coal Mining Industry', Abstract, September 2010 http://www.acarp.com.au/. Accessed 17 April 2013.

74 Gulgong's Henry Lawson Heritage Festival – June Long Weekend, brochure, 2012, p. 6.

75 Rio Tinto Coal Australia, 'Where does our Funding Go?' 'Clermont Goldfest, 'Clermont Oral and Visual History Project' – Applicant: Clermont Goldfest'.

76 L. Rickarby, 'Coal versus Hunter native title claims', *Newcastle Herald*, 24 June 201, online edition; 'Approved Bowmans [sic] Creek Diversion (Modification 6)', Yancoal, n.d., http://www.ashtoncoal.com.au/ Accessed 17 April 2013; Gloucester Coal, Duralie Coal Mine – Heritage Management Plan, March 2012, pp. 11, 14, 16; Whitehaven Coal, 'Tarrawonga Coal Mine Modification', Cultural Heritage Assessment Plan (Appendix E), http://www.whitehavencoal.com.au Accessed 17 April 2013; New Hope Coal, 'Appendix K Cultural and Heritage Report', http://www.newhopecoal.com.au

77 'Railway Corridor – Non-Indigenous Cultural Heritage Survey Report (Stage 2) – Alpha to Bowen Rail to Port Corridor', Alpha Coal

Project Environmental Impact Statement, Vol. 1, Hancock Prospecting Pty Ltd, 2010; Alpha Coal Project Environmental Impact Statement, Vol. 1, Hancock Prospecting Pty Ltd, 2010 p.15.
78   For example, see: Rio Tinto Coal Australia, 'Where does our Funding Go?' 'Western Kangoulu Keeping Place – Applicant: Hedley Johnson; Ian Kirkwood, 'Warkworth mine finds oldest Hunter artefacts', Newcastle Herald, 10 August 2012 .
79   'Newcastle Voice – Newcastle Museum Visitors Survey (Visitor Analysis and Evaluation October 2012)', Newcastle Voice & The City of Newcastle, Newcastle, October 2012.
80   H. Aston, 'Mining firm signs deal with school', Sydney Morning Herald, 3 February 2013, online edition.
81   B. Cubby, 'Festival or mineral: sponsorship tension', Sydney Morning Herald, online edition, 29 May 2009.
82   Australian Student Environment Network, 'Report-back from the Dungog Fringe Festival', 6 June 2010, http://asen.org.au/ Accessed 17 April 2013.
83   'Dungog Fringe Festival', Rising Tide Newcastle, May 2010 http://risingtide.org.au; Transition Newcastle, 'Rivers SOS Fringe Film Festival', http://www.transitionnewcastle.org.au/ Accessed 17 April 2013.
84   Z. Alcorn, 'The Herd puts green ban on coal festival', Green Left Weekly, 26 September 2009.
85   For example, Xstrata and Rio Tinto both sponsor the Upper Hunter Food and Wine Affair; Xstrata, 'Community Update', Mangoola Coal Newsletter – Issue #3, October 2009 p. 2; Rio Tinto Coal Australia, 'Where does our funding go?' 'Upper Hunter Food & Wine Affair – Applicant: Denman Chamber of Commerce'.
86   D. Page, 'King Coal's reign treads on grapes', Newcastle Herald, 28 January 2011.
87   L. Shaw, 'Bruce Tyrrell: Great Australian vineyards will outlast grubbing-up phase', Decanter.com, 29 April 2010; N. Klein, 'Winemaker Bruce Tyrrell brings his coal gas mining concerns to big smoke', The Telegraph, 9 November 2011.
88   For example, the BMA form says: 'Preference will be given to: ... Projects that provide BMA with appropriate recognition for example BMA branding on sporting club uniforms and equipment.' and 'Successful applicants are required to recognise BMA's contribution of the sponsored project [sic]. This should include any one or more of the following methods: 1) Flyers, brochures, event programs, 2) Local print media (includes school newsletters, community newspapers, flyers, brochures), 3) Launch of the event which involves attendance of BMA representatives (etc)...' BMA Community Donations, Sponsorships and Partnerships Guidelines, BMA Mitsubishi Alliance, http://www.bhpbilliton.com, Accessed 17 April 2013.
89   According to the company's annual report, Xstrata's coal division generated $2.81 billion in operating profit in 2011; 38 per cent more than in 2009. Of this, the company's Australian operations generated $2.22 billion. See: Xstrata, Xstrata Annual Report 2011, 2012, p. 54, 57, 58; For the $13.9 million figure, see: 'CSI 2012 – A Snapshot', Xstrata Coal, http://www.xstratacoal.com Accessed 17 April 2013; Xstrata, 'What Matters: Xstrata Corporate Social Involvement Program, 2012.
90   For example, Rio Tinto claimed $294 million in 'community contributions' in 2011, compared with nearly $19.6 billion in profits the previous year. See: Rio Tinto, 'Summary Income Statement', Rio Tinto Annual Review – 2011, http://www.riotinto.com; 'From the earth to the world'; Rio Tinto, Annual Review – 2011, Accessed 17 April 2013.

5 KING COAL'S MUSCLES

1   N. Williams, Speech to the Coaltrans Asia conference in Bali, June 4, 2012.
2   Australian Coal Association, 'Let's Cut Emissions Not Jobs', advertisement, September 2009.
3   See US Department of Energy, DOE's Carbon Capture Program, Department of Energy website, accessed March 2013; Congressional Budget Office, Federal Efforts to Reduce the Cost of Capturing and Storing Carbon Dioxide, Congressional Budget Office, 28 June 2012, p. 15.
4   ACIL Tasman, Economic Assessment of CPRS's Treatment of Coal Mining, May 2009.
5   B. Keane, 'Coal Association scores own-goal on emissions trading', Crikey, 29 September 2009; C. Yeates, 'Boom forecast for coal

output', *Sydney Morning Herald*, 14 September 2009. Online edition.
6   M. Wilkinson, B. Cubby & F. Duxfield, 'Ad campaign aims to crush emissions trading plan', *Sydney Morning Herald*, 7 November 2009. Online edition.
7   Minerals Council of Australia, November 2008, pp. 7–11.
8   R. Aedy, 'Michell Hooke, CEO of the Minerals Council of Australia', 'Sunday Profile', ABC Radio, 2 March 2012.
9   K. De Lacy, 'Buck stops with Rudd and the bill is due', *The Australian*, 12 June 2010, online edition; see also 'Q&A with Keith De Lacy', Company Director Magazine, 1 November 2012.
10 P. Martin, 'Mining tax: how Canberra got diddled', *Sydney Morning Herald*, 14 February 2013, online edition; J. Swan, 'Lobbyists unmasked', *Sydney Morning Herald*, 15 April 2012.
11 In the first six months of the new tax, to the end of 2012, the government collected just $126 million. W. Swan, 'Minerals Resource Rent Tax revenue', Media Release, 8 February 2013.
12 Rio Tinto, 'Taxes paid in 2012: A report on the economic contribution made by Rio Tinto to public finances', Rio Tinto, March 2013, p. 9.
13 M. Davis, 'A snip at $22m to get rid of PM', *Sydney Morning Herald*, 2 February 2011. Online edition.
14 Australian Coal Association, 'Coal campaign to highlight regional impact of carbon tax', Media Release, 18 July 2011; See also 'Carbon tax will cripple Australian coal industry: new study', Media Release, Institute of Public Affairs, June 2011. ANDEV, 'Carbon tax will cripple Australian coal industry: new study', Media release, 15 June 2011; Sid Maher, 'Carbon tax 'will cost 4000 coal jobs'', *The Australian*, 14 June 2011, online edition.
15 'Carbon tax to cost a billion in lost royalties', Media Release, Queensland Resources Council, 2011.
16 According to Climate Change minister, Greg Combet, the average impact of the carbon price on emissions from open cut coal mines in Queensland is 39 cents per tonne of coal mined. By contrast, world coking coal prices have fallen from over $US200 a tonne in June to around US$175 a tonne now. See: 'Taxes to blame for mining's demise', *Coffs Coast Advocate*, 21 September 2012.
17 T. Edis & T. Wood, New protectionism under carbon pricing, Grattan Institute, Melbourne, September 2011, p. 26.
18 D. Crowe, 'Labor must reply to scares', *Australian Financial Review*, 9 March 2011.
19 M. Chambers, 'Carbon analysis rebuffs miners', *The Australian*, 17 June 2011, online edition.
20 I. Verrender, 'Dark mills of resources industry churn out propaganda overtime', *The Age*, 22 November 2011.
21 G. Hoy, 'Blackout warnings as electricity demand rises', ABC 8 June 2011.
22 S. Maher, 'Threat of carbon tax blackouts: secret report', *The Australian*, 27 May 2011, online edition.
23 In 2012 nine power companies received cash payments amounting to $1 billion from the Energy Security Fund in 2012. Department of Climate Change and Energy Efficiency, 'Generation complexes eligible to receive Energy Security Fund cash payments', Department of Climate Change and Energy Efficiency website, accessed December 2012.
24 R. Aedy, 'Michell Hooke, CEO of the Minerals Council of Australia', 'Sunday Profile', ABC Radio, 2 March 2012.
25 A. Kirk, 'Miners launch new anti-tax ad campaign', AM, ABC Radio National, 13 April 2012. ; see also A. Kirk, 'Senator urges tax hike amid miners' ad blitz', 'The World Today', ABC Radio National, 13 April 2012; F. Anderson, 'Diesel slug looms in tax cut deal', *Australian Financial Review*, 22 March 2012, online edition; L. Tingle, 'Relief for small business, but mine tax breaks safe', *Australian Financial Review*, 14 April 2012, online edition.
26 S. Cullen, 'Miners warn Government over tax changes', ABC, 13 February 2013.
27 T. Albanese, Rio Tinto CEO, 'Mining issues, a global view', Speech at the Melbourne Mining Club in London, 8 July 2010.
28 P. Wilson, 'Rio's Albanese uses Rudd example as a warning to other governments', *The Australian*, 10 July 2010, online edition.
29 Minerals Council of Australia, 2011 *Annual Report*, p. 61.
30 See G. Pearse, *High & Dry: John Howard, climate change and the selling of Australia's future*, Viking, 2007, p. 231.
31 J. Cohen, 'The Greenhouse Mafia', 'Four

Corners', ABC TV, Monday 13 February 2006; G. Pearse, *High & Dry*, Viking, 2007, pp. 234–5.
32 From the outset the Australian Coal Association and the Business Council of Australia were funders of the project and later on coal mining companies such as BHP and CRA (a precursor of Rio Tinto) and the Electricity Supply Association of Australia paid $50 000 a year to sit on the steering committee overseeing the development of the model.
33 See for example, Alexander Downer, Minister for Foreign Affairs, to the 'Global Emissions Agreements and Australian Business' Seminar, Melbourne, 7 July 1997.
34 J. Cohen, 'The Greenhouse Mafia', 'Four Corners', ABC TV, Monday 13 February 2006.
35 A. Fowler, 'Leaked documents reveal fossil fuel influence in White Paper', PM, ABC Radio National, 7 September 2004.
36 Australian Greenhouse Office, Government Response to Tambling Renewable Energy Target (MRET) Review Recommendations', Australian Government, August 2004.
37 Prime Minister John Howard, 'Prime Ministerial Task Group on Emissions Trading', Media Release, 10 December 2006.
38 Prime Ministerial Task Group on Emissions Trading, *Report of the Task Group on Emissions Trading*, May 2007, p. 147.
39 Task Group on Emissions Trading, Department of Prime Minister and Cabinet.
40 Whitehaven Coal, 'Board of Directors', accessed February 2013.
41 T. Walker, 'Life in the afterglow', *Australian Financial Review Magazine*, 29 June 2012, p. 20, online edition.
42 Walker, 'Life in the afterglow', p. 20.
43 Cuestas Coal, *Annual Report 2012*, September 2012, p. 17.
44 'The Team', ECG Advisory Solutions website, accessed February 2013; M. Drummond, 'Costello puts on new hat at ECG Advisory', *Australian Financial Review*, 6 February 2013, online edition.
45 Another ECG client is another Victorian brown coal mining company, Mantle Mining. 'ECG Advisory Solutions', Public Sector Standards Commissioner website.
46 RFC Cambrian, 'Exergen: Investment Opportunity', RFC Cambrian, October 2012, p. 2; L. Nelson, 'Future not assured', *Latrobe Valley Express*, 13 August 2012.
47 M. Chambers, 'Palmer son elevated as float ditched and dynasty takes shape', *The Australian*, 7 April 2012, online edition; N. Evans, 'Downer joins Rinehart board', *The West Australian*, November 22, 2012, online edition.
48 Australian Government Register of Lobbyists, 'Bespoke Approach', Department of Prime Minister and Cabinet.
49 Tony Walker, 'Life in the afterglow', *Australian Financial Review Magazine*, 29 June 2012. p. 20, online edition.
50 B. Sharples & J. Scott, 'BHP, Xstrata Group Offers $4 Billion for Australia Queensland Rail Assets', Bloomberg, 26 May 2010.
51 Queensland Integrity Commission, 'Enhance Corporate Pty Ltd', Queensland Integrity Commissioner website, accessed February 2013.
52 Queensland Integrity Commission, 'Next Level Holdings Pty Ltd', Queensland Integrity Commission website, accessed February 2013.; See also A. Fraser, 'LNP guru gets Gina on lobby register', *The Australian*, July 17 2012, online edition.
53 Independent Commission Against Corruption, *Investigation into corruption risks involved in lobbying*, Independent Commission Against Corruption, November 2010, pp. 19– 20.
54 'NSW SLAPPs/NSW Minerals Council vs Rising Tide', SourceWatch, accessed February 2013.
55 Nikki Williams (CEO of Australian Coal Association), Letter to Barry O'Farrell, 20 March 2012. Stephen Galilee (Chief Executive Officer, New South Wales Mining Council), Letter to Barry O'Farrell, 9 October 2012.
56 S. Galilee, CEO of New South Wales Minerals Council, Letter to Premier Barry O'Farrell, 9 October 2012.
57 H. Aston, 'Miners lobbied premier to pull plug on environmental legal centre', *Sydney Morning Herald*, 10 January 2013, online edition.
58 M. Roche, 'Re: QUT research', email to Professor Peter Coldrake, 22 June 2011.
59 S. Maher, 'Carbon tax 'will cost 4000 coal jobs': Exclusive', *The Australian*, 14 June 2011, online edition.
60 Australian Government, 'How will coal mining companies be affected by the carbon

price?', Clean Energy Future website, accessed March 2013.
61 J. Burke, 'Gina Rinehart urges Australians to fight against carbon and mining taxes', *Australian Mining*, 5 May 2011.
62 Ten Network Holdings, 'Gina Rinehart to join Ten Holdings Board', Media Release, 26 November 2010; Ten Network Holdings, 'About Us: Georgina Rinehart', accessed February 2013.
63 C. Kruger & B. FitzGerald, 'Rinehart digs out a chunk of Channel 10', *Sydney Morning Herald*, 23 November 2010, online edition.
64 A. Bolt, 'Mining magnate Gina Rineheart stakes a claim on Channel 10', *Herald Sun*, 24 November 2010, online edition.
65 When asked by the ABC whether Rinehart had asked Lachlan Murdoch and/or James Packer to hire Bolt, her spokesman did not directly answer the question. See Gina Rinehart, 'Questions for Gina Rinehart, Answers on behalf of Hancock Prospecting Pty Ltd, 'Four Corners', ABC Television, June 2012.
66 A. Meade, 'Ten grooming Bolt for talk show debut', *The Australian*, April 05, 2011. Online edition.
67 In 2010 she helped fund Monckton's $100 000 tour to Australia and co-sponsored his 2011 tour. See G. Readfearn, 'Australia's place in the global web of climate denial', The Drum Opinion, ABC, 29 June 2011.
68 G. Readfearn, 'Monckton, Rinehart and a plan to capture the Australian media', Graham Readfearn (blog), 1 February 2012.
69 Rinehart bought her share in Farfax Media in three tranches. The first around December 2010 was for approximately 4 per cent at an estimated cost of $100 million. In February 2012 she bought a further 9.9 per cent for $192 million. See J. Chessell, N. Shoebridge & S. Thompson, 'Rinehart poised to snare 15pc of Fairfax', *Australian Financial Review*, 1 February 2010. Online edition.
70 Fairfax Media, 'Publications, Websites and Mobile Device Applications', January 2013.
71 P. Ryan, 'Singo: Gina's my mate, Fairfax Charter out-of-date', AM, ABC Radio National, 19 June 2012.
72 T. Treadgold, 'Billionaires List: Miner's Daughter', *Forbes Magazine*, 28 March 2011. Online edition.
73 J. Barrett & N. Shoebridge, 'Gina's agenda', *Australian Financial Review*, 6–7 August, 2011,
p. 46, online edition. In effect, Hancock's company wanted journalists not to be journalists at all.
74 M. Hawthorne, 'Rinehart demands notes from Fairfax journalist', *Sydney Morning Herald*, 13 March 2013, online edition; 'Rinehart subpoenas another journalist', *PerthNow*, 14 March 2013, online edition.
75 M. Attard, 'Gina Rinehart', 'Sunday Profile', ABC Radio, 5 December 2008.
76 J. Burke, 'Gina Rinehart urges Australians to fight against carbon and mining taxes', *Australian Mining*, 5 May 2011.
77 G. Rinehart, 'Resources the life-raft in an economic storm', JoNova (blog), 7 December 2011.
78 G. Rinehart, 'Questions for Gina Rinehart, Answers on behalf of Hancock Prospecting Pty Ltd, 'Four Corners', ABC Television, June 2012.
79 N. Evans, 'Gina Rinehart appoints Prof Ian Plimer to two boards', *PerthNow*, 1 February 2012, online edition.
80 P. Klinger, 'Rinehart outlines timetable for Roy Hill', *The West Australian*, 12 July 2011, online edition.
81 G. Rinehart, 'Questions for Gina Rinehart, Answers on behalf of Hancock Prospecting Pty Ltd, 'Four Corners', ABC Television, June 2012.
82 B. Burton, 'Conservative Group to Advise Gov't on Accrediting NGOs', Inter-Press Service, 9 August 2003; B. Burton, *Inside Spin: The dark underbelly of the PR industry*, Allen & Unwin, August 2007, p. 117.
83 B. Burton, *Inside Spin*, Allen & Unwin, 2007, p. 127.
84 B. Cubby & A. Lawes, 'The benefit of the doubt', *Sydney Morning Herald*, 8 May 2010, online edition.
85 Cubby and Lawes, 'The benefit of the doubt', *Sydney Morning Herald*, 8 May 2010.
86 'Abbott, Bolt, Rinehart fawn in the IPA court of King Murdoch', Crikey, April 5, 2013.
87 T. Abbott, Address to Institute of Public Affairs 70th Anniversary Dinner, Melbourne, April 5, 2013.
88 A. Moran, Inquiry into the National Access Regime: Transcript of Proceedings, Productivity Commission, May 28, 2001, p. 46.
89 A. Moran, Evidence to the Senate Economics Legislation Committee, 22 May 2009, p. 32.
90 A. Moran, 'Heading backwards with power

generation', *Herald Sun*, 20 March 2010, online edition; A. Moran, 'Taxpayers bear the cost of dumping coal', *Sydney Morning Herald*, 21 April 2011, online edition.
91 Moran, 'Heading backwards with power generation', *Herald Sun*, 2010; T. Wilson & A. Moran, 'Productivity commission shows Australia has high taxes on carbon emissions', Media Release, 28 June 2011.
92 A. Fowler, 'Against the wind', 'Four Corners', ABC TV, 11 July 2011.
93 A. Moran, 'Can wind power earn a place in Australia's energy future?', Presentation to the Australian Environment Foundation Annual Conference, 16 October 2010.
94 M. Newman, 'Against the wind: The pursuit of clean energy has relegated ordinary people to the status of collateral damage', *The Spectator*, 21 January 2012, online edition.
95 M. Newman, 'Submission to Federal Senate Inquiry into Excessive Noise from Windfarms', 31 October 2012.
96 M. Newman, 'Maurice Newman's address to ABC staff', *The Australian*, 11 March 2010, online edition.
97 M. Newman, 'Losing their religion as evidence cools off', The Australian, 5 November 2012; M. Newman, 'ABC clique in control of climate', *The Australian*, 18 December 2012, online edition.
98 'Tony Abbott visits Beaufort', Pyrenees Advocate, 2 October 2009, p. 5; K. O'Brien, 'Tony Abbott joins 'The 7.30 Report'', '7.30 Report', ABC TV, 2 February, 2010.
99 T. Abbott, 'Strengthening Australia's Productivity', Liberal Party of Australia, 2 November 2012.

6 'CLEAN COAL' RUSE

1 'Zero Carbon Australia Sydney Launch Event Video with Bob Carr and Malcolm Turnbull', Beyond Zero Emissions, September 2010, http://bze.org.au/ Accessed 25 April 2013
2 Questions without Notice – oil, Hansard, House of Representatives, 17 October 1990, pp. 3, 56.
3 Questions without Notice – coal industry: greenhouse gas emissions, Hansard, House of Representatives, 15 October 1990, pp. 2, 818.
4 'Black coal utilisation: cleaner use, bigger export dollars', Media Release, Peter Cook, Minister for Industry, 15 December 1994.
5 'Black coal goes green at new cooperative research centre', Media Release by Peter Cook, Minister for Science, 25 September 1995
6 Speech by David Beddall MP, Minister for Resources, 10th International Conference on Coal Research, Australian Coal Association, Sheraton Hotel, Brisbane 9 October 1994.
7 In 1994 Australia produced about 192 million tonnes of salable black coal. 55 million tonnes were used domestically; 136 million tonnes were exported. Since that time, total black coal production has increased to over 350 million tonnes, with exports more than doubling. Emissions from Australian coal annually have increased by over 80%. See: G. Armitage, Resources and Energy Statistics 2012, Bureau of Resources and Energy Economics, Canberra, 5 December 2012, p. 43; see also Productivity Commission, The Australian Black Coal Industry, Inquiry Report, AusInfo, Canberra, 1998.
8 J. Howard, Safeguarding the Future: Australia's Response to Climate Change, Statement by the Prime Minister of Australia, the Hon. John Howard MP, 20 November 1997.
9 W. Parer, Opening Address by Warwick Parer, Minister for Resources and Energy, Australian Coal Conference and Trade Exhibition, Jupiter's Casino – Gold Coast, 17 May 1998; See also: Productivity Commission, The Australian Black Coal Industry, Inquiry Report, AusInfo, Canberra, 1998.
10 W. Parer, Opening Address, Australian Coal Conference and Trade Exhibition, 1998
11 See for example, Factors affecting the take-up of clean coal technologies, Coal Industry Advisory Board, IEA/OECD, France 1996.
12 'The Promise of Carbon Capture and Storage', Brochure, ExxonMobil Corporation, n.d.
13 'Focus – Carbon capture and storage', World Energy Council, http://www.worldenergy. org. Accessed 26 April 2013; M. Al-Juaied & A. Whitmore, 'Realistic Costs of Carbon Capture' Discussion Paper 2009–08, Belfer Center for Science and International Affairs, Harvard University, Cambridge, July 2009.
14 'Australia backs out of climate protocol', Environmental News Service, 6 June 2002, accessed 26 April 2013.
15 Energy Task Force, Department of Prime Minister and Cabinet, *Securing Australia's Energy Future White Paper*, Commonwealth

Government, Canberra, 2004.
16 'CCS fails to deliver on promise', 'Lateline', ABC Television, 4 August 2011.
17 'Future is coal and nuclear, Howard says', *Sydney Morning Herald*, 31 January 2007, online edition.
18 see 'Additional $100 million boost to Clean Coal' Joint Media Release, Ian Macfarlane – Minister for Industry, Tourism and Resources and Malcolm Turnbull, Minister for the Environment and Water Resources, 12 March 2007.
19 W. Frew, 'Emissions policy: it's either coal or coal', *Sydney Morning Herald*, 8 March 2006; See also: 'Blueprint Number 6 – Protecting Australia from the Threat of Climate Change', Media Release by Anthony Albanese MP, 6 March 2006.
20 J. Freegard & M. Warner, '$420m to catch some sun power', *Herald Sun*, 25 October 2006.
21 O. Galacho, 'Olga Galacho: Carbon's rocky road', *Herald Sun*, 15 December 2007.
22 'Global Giant Joins Smart State Quest for Clean Coal Power', Media Statement, Peter Beattie – Queensland Premier and Minister for Trade, 16 June 2005; S. Lewis & D Shanahan, 'Howard pushes for 'new' Kyoto agreement', News.com.au, 1 November 2006.
23 D. Giles, 'Clean coal station', *Courier Mail*, 10 February 2007.
24 T. Brown & S Taylor, 'Coal', 'Sixty Minutes', Channel Nine Television, 18 February 2007.
25 'Rudd plans $1.5 billion clean coal fund', *Sydney Morning Herald*, 25 February 2007.
26 R. Garnaut, *Garnaut Climate Change Review: Interim Report*, February 2008, p. 5.
27 J. Darnbrough, 'Carbon capture critical to Australian future', *Australian Mining*, 8 September 2008.
28 For example, on 12 December 2008 the Rudd government announced that '$580 million of today's investment will be used to expand capacity along the rail corridors connecting Hunter Valley coal mines to the port of Newcastle. This $1 billion project will more than double the amount of coal being transported to export markets from 97 to 200 million tonnes a year.' See '$4.7 billion nation building package,' Media Statement by the Prime Minister, 12 December 2008. Other sources: Budget 2008–09, Budget Paper Number 2, Part 2: Expense Measures (Climate Change), Table 3 Tackling Climate Change, www.budget.gov.au/2008-09 Accessed 26 April 2013
29 D. Wells, Correspondence with Martin Ferguson – Minister for Resources and Energy, Dick Wells – Chairman, National Low Emissions Coal Council, 24 September 2010, p. 13.
30 'Press Conference with the Minister for Resources, Energy and Tourism, Martin Ferguson', Interview by Kevin Rudd, Prime Minister's Courtyard – Parliament House, Canberra, 18 September 2008.
31 'Rudd's $100m Climate Institute', *Sydney Morning Herald*, 19 September 2008.
32 Press Conference with the Minister for Resources, Energy and Tourism, Martin Ferguson'.
33 'Rudd struts world stage with global carbon capture and storage institute', *The Australian*, 10 July 2009.
34 Global Carbon Capture and Storage Institute, International Advisory Panel, Global Carbon Capture and Storage Institute, http://www.globalccsinstitute.com/. Accessed August 2012.
35 'Coal21', Climate Change Action, http://www.nswmin.com.au. Accessed 25 April 2013.
36 The newgencoal site has been edited extensively in recent years. Some old footage detailing various video contributions from researchers at CSIRO and other institutions is at http://www.thebigpugh.com/ and http://pixelnourish.com/resume/ The authors also hold copies of various video clips and online advertisements undertaken by the ACA as part of the newgencoal campaign. See also: C. Hamilton, 'Why are CSIRO scientists spruiking for the coal industry?', *Crikey*, 7 July 2009.
37 *Response to the Carbon Pollution Reduction Scheme Green Paper*, CFMEU, 10 September 2008, pp. 3, 36.
38 'Historic alliance calls for a national task force on carbon capture and storage', Joint Media Statement by the Australian Coal Association, the CFMEU, Climate Institute and WWF, Wednesday, 16 April 2008.
39 'Historic alliance calls for a national task force on carbon capture and storage', Joint Media Statement.
40 'Editorial: Nuclear report is radioactive for ALP', Editorial, *The Australian*, 23 November

2006; 'Editorial: APEC points the way on trade and power', Editorial, *The Australian*, 18 November 2006; 'Editorial: Fight just begun for Kim Beazley', Editorial, *The Australian*, 14 November 2006; 'So Bob, do you really want to save the planet?', Editorial, *The Australian*, 14 July 2011.

41 A Morton, 'Greens call for government scientist to quit Rio Tinto', *The Age*, 6 October 2003; A. Fowler, 'Questions raised over chief scientist's Rio Tinto role', '7.30 Report', ABC Television, 8 December 2003; S. Cauchi, 'Senior scientist accused of conflict', *The Age*, 7 October 2003; K. Davidson, 'The coalition of the global polluters', *The Age*, 23 October 2003.

42 For example, ABARE projected over 1000 TWh of coal and gas-fired CCS by 2030. See: 'Economic Impact of Climate Change Policy: the role of technology and economic instruments', *ABARE Research Report 06.7*, ABARE, July 2006, p.37.

43 See, for example: 'Economic Impact of Climate Change Policy: the role of technology and economic instruments', p.2–3, 8.

44 The Heat is On, Energy Futures Forum, CSIRO, 2006, pp. 99–100

45 S. Peatling, 'Coal will be the new asbestos, says Flannery', *Sydney Morning Herald*, 9 February 2007; See also: 'Now or Never: A sustainable future for Australia', *Quarterly Essay*, # 31, Black Inc, pp. 28–9.

46 For example, former head of the federal Environment Department David Borthwick became chair of the CO_2CRC ; former head of the Industry Department Russell Higgins was Chair of the Centre for Coal in Sustainable Development and later became chair of the Global CCS Institute and Chair of the CSIRO Energy Transformed Flagship Advisory Committee ; Ralph Hillman, former ambassador for the environment was also appointed as Chief Executive of the Australian Coal Association around this time.

47 For more on Siemens' role in expanding coal-fired power in developing country markets, see for example: 'Siemens increases its share in joint venture with Shanghai Electric', Media Release, Siemens, 28 June 2010; 'Siemens to triple steam turbine capacity in India', domain-b.com, 29 May 2009; 'Siemens to expand steam turbine production in India for EUR40 million' Media Release,

Siemens, 2 June 2009; and 'Siemens ready to meet surging industrial turbine demand', Jakarta Post, 18 May 2012; Re: Tata, see: 'Tata Power', Sourcewatch.org, Accessed 26 April 2013.

48 Working Group III: The IPCC Response Strategies, Intergovernmental Panel on Climate Change, 1990, p. 70.

49 Working Group II: Impacts, Adaptations and Mitigation of Climate Change: Scientific-Technical Analyses, Intergovernmental Panel on Climate Change, p. 609.

50 Stern drew extensively on the IPCC Special Report on CCS from 2005 in his report: N. Stern, *The Stern Review—the economics of climate change*, Cabinet Office – HM Treasury, January 2007, pp. 223, 238, 525, 526, 536.

51 'International Developments – CO_2 Capture and Storage & Clean Coal', Presentation by Kelly Thambimuthu, Chairman, International Energy Agency (IEA) Greenhouse Gas R&D Programme & Chief Executive Officer, Centre for Low Emission Technology, Brisbane, NSW Parliament, 7 May 2008, p. 12.

52 Carbon Dioxide Capture and Storage: A key primary abatement option for the future, International Energy Agency, 2006, p. 1.

53 *Technology Roadmap: Carbon capture and storage*, International Energy Agency/OECD, Paris, 2009 p.6; 'Tackling climate change', Australian Coal Association, http://www.newgencoal.com.au. Accessed 26 April 2013.

54 Strategic Analysis of the Global Status of Carbon Capture and Storage Report 1: Status of Carbon Capture and Storage. Projects Globally (Final Report), Global Carbon Capture and Storage Institute, Commonwealth of Australia, Canberra, 2009. pp. 4, 7–16.

55 In all, the sequestration potential of listed integrated coal related projects was 99.45 million tonnes of CO_2 per annum.

56 Globally in 2011, the World Coal Institute estimated that 6637 million tonnes of hard coal was used. At 2.8 tonnes of CO_2 per tonne of coal, this equates to around 15.5 billion tonnes of CO_2-e from coal use (excluding lignite, of which another billion tonnes were consumed). 100 million tonnes of CO_2 stored would be just over 0.5% of that generated from global coal use in 2011. With the inclusion of lignite emissions it would be less than 0.5%. World Coal Institute, International Energy Agency.

57 P. Ashworth, S. Rodriguez & A. Miller, *Case Study of ZeroGen Project*, Energy Transformed Flagship, CSIRO, Canberra, p. 6.
58 P. Beattie, 'Cleaner world will demand carbon capture and storage', *The Australian*, 23 October 2010.
59 'BP and Rio Tinto Plan Clean Coal Project for Western Australia', Media Release, BP, 21 May 2007.
60 'Minister welcomes plans for clean-coal power', Media Release by Francis Logan, WA Minister for Energy, 23 May 2007.
61 *Monash Energy Report 2006*, Anglo Coal, http://www.angloamerican.com Accessed 26 April 2013. This document features a picture of former Premier Steve Bracks looking on as the government signs an agreement with the Monash Energy proponents.
62 *Strategic Policy Framework for Near Zero Emissions from Latrobe Valley Brown Coal*, Issues Paper – Minister's Foreword, Peter Batchelor – Minister for Energy and Resources, Government of Victoria, August 2007.
63 Moomba Carbon Storage project – A project for our time and for future generations, Brochure by Santos, Origin Energy and Beach Petroleum, 2007, http://www.santos.com Accessed 26 April 2013.
64 'Funding commitment for Moomba Carbon Storage', Media Release by Santos, 20 November 2007.
65 *Federal Efforts to Reduce the Cost of Capturing and Storing Carbon Dioxide*, Congressional Budget Office, 28 June 2012, pp. 7–8.
66 Carbon Storage Taskforce 2009, *National Carbon Mapping and Infrastructure Plan – Australia: Concise Report*, Department of Resources, Energy and Tourism, Canberra. p.18.
67 Carbon Storage Taskforce 2009, *National Carbon Mapping*, p. 14.
68 'Gorgon Project (Barrow Island)', Government of Western Australia Department of State Development, Government of Western Australia, accessed July 2012.
69 *Water and the electricity generation industry - implications of use*, Report by ACIL Tasman and Evans and Peck, National Water Commission, Waterlines report No 18, August 2009.
70 See, for example, W. Montgomery & A. Smith, Price, Quantity and Technology Strategies for Climate Change Policy, CRA International, Washington DC, 2005 .
71 'CCS in the International Arena', Global Carbon Capture and Storage Institute, http://www.globalccsinstitute.com Accessed 26 April 2013; 'New UNFCCC report confirms coal power offset projects will generate millions of artificial carbon credits', Media Release, Climate Justice Now, 10 November 2011. See also N. Perry, 'Australian companies will be able to offset their emissions from coal-fired energy by financing new coal-fired energy projects', *The Conversation*, 24 October 2012.
72 M. Davis, 'Coal power in the CDM: Does it make sense?', Stockholm Environment Institute, 8 November 2011, http://www.sei-international.org. Accessed 25 April 2013.
73 Barack Obama's administration would later invest in a different project proposed by the FutureGen consortium dubbed FutureGen2. It too is yet to get off the ground. See: R. Smith & S. Power, 'After Washington pulls plug on FutureGen, clean coal hopes flicker', *Wall Street Journal*, 2 February 2008; See also: 'Department of Energy Formally Commits $1 Billion in Recovery Act Funding to FutureGen 2.0', Media Release by Steven Chu – US Energy Secretary, 28 September 2010; and http://www.futuregenalliance.org/
74 'Topics: Carbon Capture and Storage', International Energy Agency, http://www.iea.org/topics/ccs/ Accessed 25 April 2013.
75 Federal Efforts to Reduce the Cost of Capturing and Storing Carbon Dioxide, Congressional Budget Office, 28 June 2012, p. 15.
76 'Future of thermal coal cools', *Australian Financial Review*, 5 November 2012.
77 S. Lacey, 'BHP Billiton: 'Coal is going to decline. And frankly, it should'', *Reneweconomy*, 5 December 2012; 'Climate change influences BHP decision', *Australian Financial Review*, 3 December 2012.
78 CCS fails to deliver on promise, 'Lateline', ABC Television, 4 August 2011.
79 A. Fraser, 'Chief quits clean coal project, citing inaction', *The Australian*, 21 December 2010.
80 'Dr Kelly Thambimuthu says there are no large-scale CCS projects on the horizon in Australia', ABC News (online), 14 February 2012.
81 D. Wells, Correspondence with Martin Ferguson – Minister for Resources and Energy, Dick Wells – Chairman, National Low Emissions Coal Council, 26 October 2011.
82 'Flannery – I've changed my mind on carbon

capture', *Sydney Morning Herald*, 27 May 2010; A. Aikman, 'Tim Flannery backs coal seam gas and mining industry', *The Australian*, 25 October 2011; See also speech by Tim Flannery to the NSW Minerals Council, October 2011. Available at: http://media.theaustralian.com.au/

83 A. Kirk, 'Clean coal technology won't work: Opposition', AM – ABC Radio, 10 November 2009.

84 https://twitter.com/TonyAbbottMHR/ Accessed 25 April 2013.

85 A. Moran, 'Address to the Revolt Against the Carbon Tax', Occasional Paper, Institute of Public Affairs, 23 March 2011.

86 'Transcript of doorstop interview, Brisbane', Prime Minister Julia Gillard, 13 July 2011.

87 'Meet the Press – Interview with Hugh Riminton', Interview Transcript, Craig Emerson, Minister for Trade and Competitiveness, 17 June 2012.

88 '(Attachment 2: Paying for Recovery and Reconstruction) Rebuilding after the floods', Media Release by Prime Minister Julia Gillard, 27 January 2011.

89 P. Beattie, 'Cleaner world will demand carbon capture and storage', *The Australian*, 23 October 2010.

90 'Zero Carbon Australia Sydney Launch Event Video with Bob Carr and Malcolm Turnbull', Beyond Zero Emissions, September 2010, http://bze.org.au/ (Accessed 25 April 2013).

91 A. Morton, 'Coming Clean', *The Age*, 4 November 2009.

92 L. Taylor, 'Coal hard light of day for dud scheme', *Sydney Morning Herald*, 17 June 2012; L. Taylor, 'Waste of energy or fine chance for a ballroom blitz?', *Sydney Morning Herald*, 17 June 2012.

93 *The Status of CCS Projects Interim Report 2010*, Global Carbon Capture and Storage Institute. p. 2.

94 'Global CCS Institute e-News Update', August 2009 http://cdn.globalccsinstitute.com Accessed 25 April 2013.

95 See: Carbon capture and storage – mobilising private sector finance, The Climate Group, Ecofin Research Foundation and Global Change Institute, undated; R. Posner, 'Global climate focus back in New York with Climate Week', Climate Week NYC Blog, The Climate Group, 25 July 2011; R. Posner, 'Small step forward for CCS in the US', Insights: Perspectives from around the world, Global Carbon Capture and Storage Institute website 24 August 2011 http://www.globalccsinstitute.com. Accessed 26 April 2013.

96 Global Carbon Capture and Storage Institute, 2009–10 Second progress report, n.d., p. 19.

97 J. Connor, 'Why CCS shouldn't be viewed as the unpopular cousin', *Climate Spectator*, 5 December 2012.

98 M. Ferguson, Correspondence with Prime Minister Kevin Rudd, Minister for Energy and Resources, 16 December 2009, p. 2.

99 L. Taylor, 'Coal hard light of day for dud scheme', *Sydney Morning Herald*, 17 June 2012.

100 'Address to the Hunter Business Chamber's Mining Industry Luncheon', Speech by Greg Combet - Minister for Climate Change and Energy Efficiency, The Glades and Wedding and Conference Centre, Warners Bay, 20 May 2011.

101 Between the South West Hub and the CarbonNet proposals, for example, the total amount of CO_2 captured annually (according to their proponents) is 3–28mt per annum. New mines proposed in the Galilee Basin would add 700 million tonnes of CO_2 annually. See: http://www.globalccsinstitute.com/projects; and *Cooking the climate: wrecking the reef – the global impact of coal exports from Australia's Galilee Basin*, Greenpeace Australia Pacific, September 2012.

102 *Power Generation from Coal Measuring and Reporting Efficiency Performance and CO_2 Emissions*, IEA Coal Industry Advisory Board (CIAB), 2010. p. 59.

103 As the IEA notes, 'More advanced coal technologies are being deployed, but inefficient coal technologies still account for almost half of new coal fired power plants being built' See: 'Higher-efficiency and lower-emission coal', International Energy Agency, Energy Technology Perspectives 2012 (Technology Overview Notes – Coal), http://www.iea.org Accessed 25 April 2013.

104 Power Generation from Coal Measuring and Reporting Efficiency Performance and CO_2 Emissions, IEA Coal Industry Advisory Board (CIAB), 2010. p. 57.

105 See: Figure 1. CO_2 emissions by fuel, CO_2 Emissions from Fuel Combustion -- Highlights (2012 edition), International Energy Agency, 2012. p. 124.

106 The Kogan Creek Solar Boost project will save an estimated 35,600 tonnes of CO_2

annually once it is up and running. The 750MW coal fired power station to which it is attached generates over 3.6 million tonnes of CO_2 annually. So the Solar Boost will save about 0.97% of the emissions from the power station. See: 'Kogan Creek Solar Boost Project', Fact Sheet, CS Energy, August 2012; and http://carma.org/plant/detail/22602

107 For example, see: '$20 million investment in advanced biofuels', Media Release by Martin Ferguson – Minister for Energy and Resources, 24 February 2012; See also: G. Pearse, 'Coal's Next Alibi - The Coal Industry's Coal-fed Algae Plan, *The Monthly*, August 2011

108 'Partnering with MBD Energy to reduce greenhouse gas emissions', Sustainable Development, Anglo American, http://www.angloamerican.com.au. Accessed 26 April 2013

109 See: M. Rasini, 'Coalition hails JCU algae aim', *Townsville Bulletin*, 18 May 2011; T. Edis, 'Climate Spectator: Coalition's RET affirmation', *Business Spectator*, 28 September 2012

110 'Abbott Doorstop - MDB Energy Algal Research and Development Facility, Townsville', Transcript, Liberal Party of Australia, 10 December 2009.

111 'Algae link to Australia's clean coal technology', http://news.xinhuanet.com 16 October 2009; P. Beattie, 'Nation must jump on biotech wagon', *The Australian*, 25 June 2011

112 *Accelerating the uptake of CCS: Industrial Use of Captured Carbon Dioxide*, Parsons Brinckerhoff in collaboration with the Global CCS Institute, pp. 41, 43, 63–4.

113 According to the US DOE, 'The CO_2 generated by the power plant can only be effectively used by the algae during the photosynthetically active sunlight hours. As a result, the greenhouse gas emissions offset will be limited to an estimated 20 per cent to 30 per cent of the total power plant emissions due to CO_2 off-gassing during non-sunlight hours and the unavoidable parasitic losses of algae production', National Algal Biofuels Technology Roadmap, Biomass Program, US Department of Energy, May 2010, p. 80.

114 According to the IEA, global CO_2 emissions from coal were around 6 Gt in the mid-1980s. By 2010 they were around 13Gt. For more detail see: *Emissions from Fuel Combustion – Highlights* (2012 edition), International Energy Agency, 2012 f p. 8.

7 OVERTHROWING BIG COAL

1 M. Lucas, Cited in 'Bloomberg at ARPA-E: 'Coal is a Dead Man Walking', Berkley Energy and Resources Collaborative, March 6, 2013.

2 W. Steffen, The Angry Summer, Climate Commission, March 2013, http://climatecommission.gov.au/wp-content/uploads/130408-Angry-Summer-report.pdf

3 See David Spratt, 4 Degrees Hotter, Climate Action Centre, February 2011; P. Whetton & D. Karoly, Australian climate at four degrees or more of global warming, Proceedings of Four Degrees of More?: Australia in a hot world conference, July 2011.

4 S. Barley, N. Hawtin, C. Brahic & T. Simonite, 'No rainforest, no monsoon: get ready for a warmer world', *New Scientist*, 30 September 2009, online edition.

5 Potsdam Institute for Climate Impact Research and Climate Analytics (PICIRCA), *Turn down the heat: why a 4°C warmer world must be avoided*, World Bank, November 2012, p. xii.

6 PICIRCA, Turn down the heat, World Bank, November 2012, p. xiii.

7 PICIRCA, Turn down the heat, World Bank, November 2012, p. ix.

8 International Energy Agency, 'Climate change', accessed February 2013. Available at: http://www.iea.org/topics/climatechange/

9 L. Johnson, Partner, Sustainability and Climate Change PWC, 'Foreword' in *Too late for two degrees?: Low carbon economy index 2012*, November 2012, p. 1.

10 J. Hansen et al., 'Target atmospheric CO_2: Where should humanity aim?', Journal of Atmospheric Science, vol. 2, October 2008, pp. 217–23.

11 B. Cubby, 'New kinds of power to the people', *Sydney Morning Herald*, 18 April 2013.

12 J. Gordon & R. Millar, 'Coal safe come 'hell, high water', *The Age*, 4 March 2011, online edition.

13 K. Barlow, 'Thousands flee record floods', 'Lateline', ABC TV, 29 January, 2013; See also Graham Readfearn, 'Give us a break on the climate science denial', Graham Readfearn (Blog), 30 January 2013.

14 P. Beattie, 'Goodwill and a concerted national

effort needed', *The Australian*, 15 January 2011, online edition.
15 S. Peatling & R. Browne, 'Queensland puts coal before coral', *Sydney Morning Herald*, 3 June 2012, online edition.
16 J. Garnaut, 'Time for change: China flags peak in coal usage', *The Age*, 6 February 2013.
17 F. Green & R. Finighan, 'Laggard to leader: How Australia can lead the world to zero carbon prosperity', Beyond Zero Emissions, July 2012.
18 Department of Climate Change and Energy Efficiency (DCCEE), Australian National Greenhouse Accounts, Quarterly Update of Australia's National Greenhouse Gas Inventory, September Quarter 2012, DCCEE, February 2013, p. 5.
19 DCCEE, Australian National Greenhouse Accounts, September Quarter 2012, February 2013, p. 6.
20 A. Pears, *Imagining Australia's Energy Services Future*, self-published, January 2006.
21 H. Saddler & G. Anderson, 'National Electricity Market emissions fall to their lowest level for ten years', Carbon Emissions Index, Pitt & Sherry, November 2012.
22 Delta Electricity, 'Munmorah Power Station to close after 45 years of operation', Media Release, 3 July 2012; Stanwell Corporation, 'Swanbank B Power Station fires up for the last time', Media Release, May 21, 2012.
23 Stanwell Corporation, 'Stanwell to withdraw Torong Power Station from service', Media Release, 11 October 2012.
24 T. Walsh & L. Ashworth, 'One of two units at Wallerawang Power Station mothballed for twelve months', *Lithgow Mercury*, 8 January 2013, online edition.
25 M. Ferguson, G. Combet & S. Crean, 'Briquette Restructuring Package', Media Release, 29 June 2012.
26 Australian Energy Market Operator, *South Australian Electricity Report*, August 2012, Table 1. Available at: http://www.aemo.com.au/Electricity/Planning/South-Australian-Advisory-Functions/South-Australian-Electricity-Report
27 Government of Western Australia, Strategic Energy Initiative Energy 2031: Building the Pathways for Western Australia's Energy Future, August 2012, p. 15. Available at: http://www.energy.wa.gov.au/1/3315/3312/cleaner_energy.pm

28 T. Arup, 'Gas-fired power plant put on hold', *The Age*, 27 December 2012.
29 B. Keane, 'Aluminium smelting: the best bang for your fossil-fuel subsidy buck', Crikey, 10 March 2011.
30 J. Greber & M. Priest, 'Labor caves in to smelter', *Australian Financial Review*, 25 June 2012; T. Arup & J. Gordon, 'State avoids $400m bill on carbon', *The Age*, 12 February 2013, online edition.
31 Bureau of Resources and Energy Economics, Australian Energy Projections, December 2012, pp. 40–1, http://www.bree.gov.au/publications/aep.html
32 Domestic coal use in the steel and cement industries is relatively minor. In 2009–10, the most recent year data is available for, the steel industry consumed 4.5 million tonnes of coal. In 2005–06, the most recent year data is available for, the cement industry consumed less than 1 million tonnes and other consumers – such as in food processing plants – another 4.9 million tonnes. Bureau of Resources and Energy Economics, *Resources and Energy Statistics 2012*, December 2012, p. 45.
33 One analyst estimates that using currently available technology Australian household energy consumption could be halved. See T. Edis, 'Driving energy demand back 30 years', *Climate Spectator*, 8 March, 2013. See also A. Pears, 'Four years of falling electricity demand: can this continue?', *The Conversation*, 21 January, 2013, online. A. Pears, 'Energy revolution or bloody war?, Reneweconomy, 20 April, 2013, online.
34 J. Freed, 'Future of thermal coal cools', *Australian Financial Review*, 5 November 2012, online edition.
35 International Energy Agency, *World Energy Outlook 2012*, November 2012, p. 157.
36 Bureau of Resources and Energy Economics, Australian Energy Projections, December 2012, p. 53.
37 New South Wales Minerals Council, 'News', Twitter, 11 February 2013.
38 Australian Coal Association, Twitter, 14 February 2013.
39 A. Grigg, 'China announces plans for a carbon tax', *Australian Financial Review*, 20 February 2013. Online edition.
40 Bureau of Resources and Energy Economics, *Resources and Energy Quarterly*, March 2013, pp. 39–40.

41 J. Garnaut, 'Time for change: China flags peak in coal usage', *The Age*, 6 February 2013, online edition.
42 J. Ma, 'Big Bang measures to fight air pollution', Deutsche Bank Markets Research, 28 February 2013, p. 4.
43 A. Grigg, 'China announces plans for a carbon tax', *Australian Financial Review*, 20 February 2013, online edition.
44 Ma, 'Big Bang measures to fight air pollution', 2013; Vishal Shah, 'DB Chinese Energy Policy Proposal: Cleantech implications', Deutsche Bank Markets Research, 1 March 2013.
45 See L. Fang, *The Myth of China's Endless Coal Demand: A missing Market for US Exports*, Greenpeace, February 2013.
46 D. Stanway & R. Lian, 'New rules may ease China pollution, won't solve steel overcapacity', Reuters, 10 March 2013.
47 World Coal Association website, 'Iron & steel', accessed March 2013.
48 Midrex, 2011 *World Direct Reduction Iron Statistics*, p. 3.
49 One emerging Australian technology uses waste rubber from tyres and plastics to reduce the need for coal and coke in electric arc furnaces. M. Nadin, 'The Innovation Challenge winner could change steel-making forever', *The Australian*, 12 December 2010, online edition; University of NSW, 'Scientia Professor Veena Sahajwalla', University of NSW website, accessed February 2013. The International Energy Agency has also flagged technical developments such as the increased use of pulverised coal and the reuse of waste gases to reduce the amount of coal needed in conventional blast furnaces. International Energy Agency, World Energy Outlook 2010, International Energy Agency, November 2012, p. 342.
50 'Proposed coal plants in India' CoalSwarm website, accessed June 2012.
51 'Thermal Power Plants on the Anvil', Prayas Energy Group, August 2011.
52 'Coal pollution causes 70,000 deaths a year in India: IMF chief', Firstpost, 12 June 2012.
53 J. Vidal, 'Indian coal power plants 'kill 120,000 people a year', says Greenpeace', *Guardian*, 10 March 2013, online edition.
54 'Tata's Mundra plant fully operational', *Business Standards (India)*, 7 March 2013, online edition.
55 A. Sasi, 'RBI to banks: Take call on stressed sector exposure', *The Indian Express*, 3 February 2012.
56 E. O'Carroll, 'World Bank backs massive India coal plant, calls it 'clean development', *Christian Science Monitor*, 9 April 2008, online edition; See Table IV.i in A. Yang & Y. Cui, 'Global Coal Risk Assessment: Data Analysis and Market Research', World Resources Institute, December 2012, p. 18.
57 E. de Place & P. MacRae, 'Coal Exports From Canada: Why coal planned for ports in Oregon and Washington cannot divert to British Columbia', Sightline Institute, July 2012.
58 JATAM, London Mining Network and Nostromo Research, *Indonesia's Coal: local impacts – global links*, Down to Earth, August 2010; See also H. Widhiarto. 'No checks on coal exploitation', *Jakarta Post*, 30 April 2012, online edition.
59 International Energy Agency, Coal: Medium Term Market Report 2012: Market trends and projections to 2017, International Energy Agency, p. 111.
60 M. Drummond, 'Upending the coal-climate change nexus', *Australian Financial Review*, January 15, 2013, online edition.
61 BHP Billiton, 'Production to cease at Gregory open-cut operation', Media Release, 10 September 2012.
62 BHP Billiton, 'Production to cease at Gregory ...', Media Release, 10 September 2012.
63 Rio Tinto, 'Blair Athol Mine to finish production', Media Release, 8 August 2012.
64 Whitehaven Coal, 'Whitehaven Coal Suspends operations at Sunnyside Mine indefinitely', Media Release, 25th October 2012.
65 Australian Bureau of Statistics, Labour Force, Australia, Detailed, Quarterly, November 2012, Table 6, Cat. No. 6291.0.55.003.
66 S. Davidson & A. de Silva, in a consultancy report for the Minerals Council of Australia, argued that the employment multiplier for the coal industry is 6.5. See S. Davidson and A. de Silva, *Costing of The Greens Economic Policies: Mining*, Minerals Council of Australia, July 2011, p. 24.
67 M. Whop, 'Norwich Park closure impacts Dysart community', ABC News, 11 May 2012; Tara Miko, 'Norwich Park closure hits Dysart', CQ News, 16 May 2012; T. Miko, 'Mine closure leaves Dysart broken', CQ News, 27 July 2012.

68 J. Freed, 'Yancoal digs in and cuts costs', *Australian Financial Review*, 28 July 2012, online edition.
69 For example, over four-fifths of coal from NSW was destined for power stations with nearly all accounted for by just four countries: Japan, China, Korea and Taiwan. See NSW Minerals Council, *NSW Mining 2012: A snapshot*, New South Wales Minerals Council, March 2013, p. 8.
70 K. Rudd interview with Paul Cleary, cited in his book, *Too Much Luck: the Mining Boom and Australia's Future*, Black Inc. Collingwood, 2011, p. 144.
71 R. Denniss & M. Grudnoff, *Too Much of a Good Thing? The macroeconomic case for slowing down the mining boom*, The Australia Institute, 2012.
72 ABS, Labour Force, Australia, Detailed, Quarterly, Table 06, 'Employed Persons by Industry', March 2013. (Cat. No. 6291.0.55.003)
73 ABS, Labour Force, Australia, Detailed, Quarterly, Table 06, 'Employed Persons by Industry', February 2013. (Cat. No. 6291.0.55.003)
74 Australian Dairy Association website, 'The Australian Dairy Industry', accessed December 2012.
75 Economics Legislation Committee – Estimates – Treasury Portfolio, Senate Hansard, 16 February 2012, pp. 25–6.
76 A. de Kretser, 'Australian coalmines under a cloud: Rio', *Australian Financial Review*, 30 November 2012, online edition.
77 Rio Tinto, Rio Tinto investor seminar 2012: Presentation, 29 November 2012, p. 34.
78 B. Chapman & K. Lounkaew, How many jobs is 23,510, really? Recasting the mining job loss, The Australia Institute, Technical Brief No 9, June 2011, p. 16.
79 Australia Institute, An analysis of the economic impacts of the China First mine, December 2011, p. 4; Australia Institute, 'Clive Palmer's new QLD mine to hit Victorian and South Australian manufacturing', Media Release, 16 December, 2011.
80 J. McCarthy, 'Dusty mine objects to tourism plan', *Newcastle Herald*, 10 February 2013, online edition.
81 The Australian Parliament set the carbon benchmark price at $23 for each tonne of greenhouse gas emissions for the first three years. Each exported tonne of Australian black coal generates approximately 2.7 tonnes of greenhouse gas emissions when burnt, and Australia exported approximately 320 million tonnes in 2012 so the approximate climate liability from exported coal is $20 billion. While some of the countries Australia exports coal to have a carbon tax – albeit as a lesser rate – some don't.
82 Department of Foreign Affairs and Trade, *Analysis of Australia's Education exports*, March 2011, p. 1.
83 E. Connolly & D. Orsmond, *The Mining Industry: From Bust to Boom*, Reserve Bank of Australia, December 2011, p. 38.
84 D. Richardson & R. Denniss, *Mining the Truth: the rhetoric and reality of the commodities boom*, The Australia Institute, September 2011, p. 23.
85 See NSW Government, Budget Paper No. 2: 2012–13 Budget Statement, Chapter 5, p. 8; Queensland Government, Budget Strategy and Outlook 2012–13, p. 48.
86 The Budget papers state that there was 'a $502 million (or 14.6 per cent) decrease in royalty and land rent revenue, primarily associated with significantly lower than expected coal exports, due to a slower recovery from 2010–11 flooding and industrial action, and the appreciation of the AU$-US$ exchange rate.' See *Queensland Government, Budget Strategy and Outlook 2012–13*, p. 48.
87 NSW Government, *2012–13 Half-Yearly Review*, December 2012, p. 1.
88 World Bank and Queensland Reconstruction Authority, Queensland Recovery and Reconstruction in the Aftermath of the 2010/2011 Flood Events and Cyclone Yasi, World Bank, World Bank and Queensland Reconstruction Authority, June 2011, p. 11.
89 'Carbon tax wil be own goal: Abbott', *Sydney Morning Herald*, 9 June, 2011; T. Abbott, Joint Doorstop Interview with Mr Ken O'Dowd MHR, Gladstone, Liberal Party of Australia, 11 March, 2011; T. Abbott, 'Joint Doorstop Interview With Senator Barnaby Joyce', Liberal Party of Australia, 12 July, 2011.
90 D. Beiuk, 'Sky yet to fall in, Tony Abbott concedes', *Sydney Morning Herald*, 25 August 2012, online edition.
91 S. Koukoulas, 'Abbott's cotton-wool wrecking ball', *Business Spectator*, 7 March 2013, online edition.
92 G. Pearse, 'Quarry Vision: coal, climate change and the end of the resources boom',

Quarterly Essay, Issue 33, 2008, p. 88.
93 'Margaret River coal mining applications terminated', ABC News, 24 July 2012.
94 N. Bita, 'Campbell Newman slams farm gate shut on miners', The Australian, 29 March 2012, online edition.
95 Mantle Mining, 'Deans Marsh Update', Mantle Mining, 28 July 2011.
96 M. Ferguson, 'Funding not Proceeding for Dual-Gas Project', Media Release, 27 July 2012; Royce Millar & Adam Morton, 'Big banks 'no' to coal plant', The Age, 21 May, 2011, online edition.
97 Sean Nicholls, 'Dead in the water: O'Farrell buries coal seam gas plans', Sydney Morning Herald, 19 February 2013, online edition.
98 B. Cubby & S. Rigney, 'Tiny Bulga wins day against mining Goliath', Sydney Morning Herald, 16 April 2013, online edition.
99 'Newcastle coal terminal plans delayed', Sydney Morning Herald, 10 April, 2013, online edition.
100 D. Claughton, 'The politics of mining in NSW', ABC Rural, 11 February 2011.
101 Essential Research, Essential Report: 21 July 2008.
102 'Phasing out Australia's coal industry', EMC, July 6.

INDEX

ABARES 136–37, 168
Abbot Point Coal Terminal, Qld 55, 70, 92
Abbott, Tony 59, 151–52, 154–55, 182, 192, 215
ABC 154
Aboriginal sites 34, 39, 123, 124
Abramovich, Roman 79
ACA *see* Australian Coal Association (ACA)
acidity of oceans 57
ACIL Tasman 129–30, 132, 146–47
Acland, Qld 49, 50–51, 123, 217
Adani (company) 5, 79
Adani, Gautam 55, 90–92, 93
Adelaide, SA 17, 32
Advanced Lignite Demonstration Program 142
advertising campaigns 100–103, 126, 128–29, 144
AEC group 56
Africa 87–88
AGL *see* Australian Gas Light Co.
Agricultural Bank of China 86
agriculture 6, 31, 33, 37–38, 43–44, 50–52, 56, 109 *see also* farmers
AIGN 135, 136–38, 139, 140
air pollution 26–29, 49–50, 197, 198, 205–7, 214
Alaska 89
Albanese, Tom 62, 88, 134
Alberta, Canada 89, 187
algae 191–93
allergies 28
Alpha, Qld 1, 112, 114–15
Alpha Coal 55
Alpha mine, Qld 5, 55, 70, 95
Alpha North mine, Qld 72
Alpha West mine, Qld 70
aluminium production 202
Ambre Energy 109
ANDEV 71–72
Andrews, Peter and Stuart 43

Anglo American 5, 30, 61, 62, 64, 66, 67, 112, 118, 172, 213–14
Anvil Hill mine, NSW 37, 40
Appin, NSW 21
appliances, electrical 25–26
aquifers 45
Arrow Energy 84
Artic ice sheet 7, 195–96
Ashton, NSW 37
Ashton Mining 38, 39
Asian Development Bank 178, 186
Asia Pacific Partnership on Clean Development and Climate 161
Association of Mining and Exploration Companies (AMEC) 72
asthma 27, 28, 107
Aston Resources 76, 77, 84
Atkinson, John 86
Aurizon 94, 106, 107, 116, 123
Austrade 87
Australia Institute 213
Australian Broadcasting Corporation 154
Australian Bureau of Agricultural and Resource Economics and Sciences (ABARES) 136–37, 168
Australian Capital Territory 203
Australian Coal Association (ACA) 102, 128, 135, 137, 144, 146, 196, 205, 213
Australian Coal Association Research Program (ACARP) 121–22
Australian dollar 6, 56, 201, 210, 211, 213, 216
Australian Environment Foundation (AEF) 153
Australian Gas Light Co. (AGL) 17, 18, 153
Australian Industry Greenhouse Network (AIGN) 135, 136–38, 139, 140
Australian Mining Industry Council 135 *later* Minerals Council of Australia
Australian National Low Emissions Coal Research and Development 165

Australians for Northern Development and Economic Vision (ANDEV) 71–72
Australian Society of Soil Scientists 51
Australian Stock Exchange (ASX) 61–62, 63, 76, 84, 216
autism 107

Bacchus Marsh, Vic 58
Baillieu, Ted 5–8
Bain, Robyn 138
Baker & McKenzie 86
Bandanna Coal 84–85
Bandanna Energy 143
banks 63–65, 198, 199
BankTrack 64
Barangaroo development, NSW 19
Barclays 64
Barlow, Jeremy 84–85
Barlow Jonker 84
Barnett, Colin 217
barons, coal 60–99
Barry, Craig 28
basalt 44
Bass (explorer) 13
Batchelor, Peter 173
Batey, Lance 42
Batterham, Robin 168, 192
BCA 135, 137, 139
Beattie, Peter 163, 167, 172, 184, 192, 197
Beazley, Kim 162
Beddall, David 157
Beijing, China 205
Bellona 186
benzene 19, 47
Bespoke Approach 142
Beutel, Glen 50
Beylando Council, Qld 108, 115
BHP 5, 30, 44–45, 54, 61–62, 97, 101, 107, 137
BHP Billiton 37, 64, 110, 122, 131, 138, 139, 150, 180–81 *see also* BHP
BHP Mitsubishi Alliance (BMA) 112, 113, 118, 120, 123, 210
Bigge, Commissioner 14
Bimblebox Nature Reserve 1–2, 3
bio-carbon capture and storage 85
biodiversity 110
birth defects 28
Bishop, Julie 95
Bjelke-Petersen, Flo 74
Bjelke-Petersen, Joh 69, 73–74
Bjelke-Petersen government 54
black coal 12–13, 29, 156 *see also* exports, black coal

'black lung' 22–24
blackouts 133, 207
BlackRock 65
Blackwater International Coal Centre (BICC) 123
Blackwood Coal 141
Blair Athol Coal mine, Qld 108, 210–11
Bligh, Anna 48, 163
Bloomberg, Mike 194
BMA *see* BHP Mitsubishi Alliance
Boao Forum for Asia 150
Bolkus, Nick 142
Bolt, Andrew 147–48, 151
Bolt Report, The 148
Bond, Peter 60, 62, 78–80, 104
Bourne, Greg 167
Bowen Basin, Qld 28, 46, 53, 73, 79, 80, 81, 82, 84, 115
Bowman, Wendy 37–38, 42
Bowman's Creek, NSW 38–39
Boyce, Greg 100, 119
BP 170, 172
Brisbane, Qld 17, 19, 60
British Gas 119
Broken Hill, NSW 41, 61, 62
brown coal (lignite) 12, 28, 58–59, 142, 172, 173, 196, 202
Brumby, John 162
BRW Rich List 66, 67
Bryson, George 23
Bulga, NSW 218
Bulli, NSW 20
Bundaberg, Qld 195
Burbank, John 88
Bureau of Resources and Energy Economics (BREE) 202, 204, 205, 206
Burke, Tony 95
Bush, George W 160, 161
bushfires 32, 194–95, 196
Business Council of Australia (BCA) 135, 137, 139
Byerwen 81
Bylong Valley, NSW 42–43

Cairns, Qld 56
Callatoota Estate 40
Camberwell, NSW 37, 38
Cameron, Anne 52
cancer 26, 27, 28
Cape Preston, Vic 14
Cape York 59
carbon capture and recycling (CCR) 191–93
carbon capture and storage (CCS) 102–3, 129, 138, 157, 158–93, 216 *see also* bio-carbon

capture and storage
Carbon Capture and Storage Flagships Program 164
carbon credits 92, 94, 178–79, 203
carbon dioxide (CO_2)
   emissions 31, 57, 71, 80, 85, 91–92, 94, 96–97, 190
   recycling 191–93
   storage 158–62, 169, 171–76, 179, 191
Carbon Mitigation Initiative 170
CarbonNet 188–89
Carbon Pollution Reduction Scheme 130
carbon price 10, 159–60, 174–75, 178, 179, 182, 192–93, 204
Carbon Sequestration Leadership Forum 161
Carbon Storage Task Force 175
carbon tax
   Australia 8, 87, 126, 130, 132, 133, 146, 152, 153, 175, 202, 214, 215–16
   China 206, 214
   Japan 130, 214
Carmichael Mine, Qld 55, 91
Caroona, NSW 44–45
Carr, Bob 156, 162, 184
Carter, Bob 151
Cascade Coal 43, 82–83
Cassoni, Paola 1–2, 3
Caterpillar 86
Cattle, Stephen 44
cattle industry 109, 110
Caval Ridge mine, Qld 54
CCS *see* Carbon Capture and Storage
Centennial Coal 40, 61, 109
Central Queensland 21, 36, 112, 116 *see also* Galilee Basin
Centre for Coal Seam Gas (CCSG) 119–20
Centre for Independent Studies 153
Centre for Low Emissions Technology 163
Centre for Sustainable Mining Practices 120
CEOs 62
Cessnock, NSW 41, 108, 115
CFMEU 146, 147
CH4 (company) 84
Channel Seven 87
Channel Ten 104, 147–48, 151, 218–19
Cheney, Dick 160–61
Chevron 176
Chifley, Ben 23
childhood development 27
China 6, 9, 29, 30, 73, 80, 86, 87, 95–97, 162, 168, 189–90, 198, 205–7, 216
China Eximbank 73
China First project 1–2, 55–56, 72, 73, 142, 213

China Investment Corporation (CIC) 65
China Power International 73
China Stone Project 96–97
Chinchilla, Qld 79, 110
Chong, Sam 82, 109
cigarettes 219
Citicorp 63, 64–65
CITIC Pacific Mining 72
'clean coal' 139, 156, 160–63, 172–73, 184, 188–93
Clean Energy Finance Corporation 92, 183, 192
Clean Energy Fund 152
'Clean Energy Future' plan 94
Clermont, Qld 108, 112, 122–23
climate change x-xi, 3–4, 7–8, 31–33, 59, 65, 71, 75–76, 83–84, 124, 127–29, 134, 147, 156, 158, 165, 188, 195–96
climate change sceptics 31, 60, 71, 139, 147–48, 150–51, 152, 154, 204, 217
Climate Commission 182, 195
Climate Group, The 186
Climate Institute 167, 187
Clinton Foundation 186
CO2 Cooperative Research Centre 161, 181
Coal21 Fund 166
Coal & Allied 112, 116
Coal and the Commonwealth 118–19
coal barons 60–99
coal demand 198–99, 218
coal dust 22–24
coal exports *see under* exports
coal-fired power plants *see* electricity, coal-fired
coal gas (town gas) 13, 16–19 *see also* coal seam gas
Coal India 93–94
Coal Industry Advisory Board (CIAB) 158, 171
Coal Mines Regulation Bill (1894) 20–21
coal ports *see* ports, coal
coal prices 30, 68, 132, 210, 211, 215
Coal River (*later* Newcastle), NSW 13–14
Coal's Assault on Human Health (Physicians for Social Responsibility) 26
coal seam gas (methane) 20, 21, 40–41, 48, 84, 110, 116, 119–20, 217–18
coal tar 16, 18, 19
Coal to Coast Festival 125
coal-washing 78
Coalworks 45
Cockatoo Coal 85, 86
'co-firing' 190
coking coal 9, 10, 29, 30, 68, 81, 87, 89, 96, 132
Collie, WA 13, 203

INDEX   249

Collins House 62
Combet, Greg 188
company directors 63
Connor, John 167
Construction, Forestry, Mining and Energy Union (CFMEU) 146, 167
consulting by universities 118–19
convicts, mining by 13–14
Cook, Peter 156, 181
Coolimba CCS power station 179
Cooperative Decarbonisation 199
Co-operative Research Centre for Mining 121
Coppabella 81
copper 19, 38
corruption 47–48 *see also* ICAC
Costa, Michael 218
Costello, Peter 141–42
cost of living 41
Cougar Energy 47
CRA 137
creeks 37–39 *see also* streamflow
Crosby Textor 130
CS Energy 190
CSIRO 7, 32, 103, 121, 166, 168, 185
Cuesta Coal 141
cultural heritage 122–24
Cummings, Bart 88
curriculums 121, 124
cyanide 18, 19
Cyclone Yasi 215

dairy farming 37, 50
Daley, John 139
Dalrymple Bay Coal Terminal 125
Darling, Harold 61
Darling Basin 176
Darling Downs, Qld 51
Darling Harbour, NSW 16, 19
Darling River, NSW 43–44
Dart Energy 84
Darwin, NT 32
Deans Marsh, Vic 217
De Beers 66
decarbonisation 196, 199–200
de Lacey, Keith 131
Denman, NSW 40
Department of Climate Change 140, 152
Derby, WA 59
Deutsche Bank 64, 206
developing countries *see* Africa; China; India
Devlin, Pat 51
Diamond, Marion 119
Dickie, Geoff 97
DIDO (drive in, drive out) workforces 53, 101, 111–13
Diesel Dash, The 79
diesel fuel rebate 133
Docklands, Vic 18–19
'dongas' 53
Downer, Alexander 142, 162
drag lines 30, 35, 86
Drake-Brockman, Judge 21
Drayton mine 214
dredging 56–57
droughts 7, 32, 197
Dublin, Ireland 26–27
Duddy, Tim 45
Duncan, Travers 82–83
Dungog, NSW 124–25
Dunk Island, Qld 104
dust disease 22–24
dust monitors 28, 49–50

Ebenezer, Qld 49
ECG Advisory Solutions 141–42
economic studies 145–46
economy
    Australian 31, 56, 97–99, 200, 211–13, 216
    in mining towns 41, 213
eczema 28
Edison, Thomas 24
editorial independence 149
education 6, 31, 118–22, 124
Elder, Jim 142
electric arc furnaces 207
electricity
    coal-fired 4–5, 12, 13, 18, 24–29, 55, 58, 90–94, 96–97, 139, 168–69, 189–90, 200–4
    demand for 177, 201
    introduction of 24–25
    renewable 137, 138, 139, 140, 153–54, 177, 190, 202–4
Emerson, Craig 183
emissions reduction target 140
emissions trading scheme 8, 119, 128–30, 139
employment *see* jobs
EnergyAustralia 133
Energy Brix power station 201
energy efficiency 203
Energy Futures Forum 168
'energy security' fund 133, 202
Energy Supply Association of Australia 135, 137, 166, 188
Energy Task Force (USA) 161
Energy White Paper (2004) 161

Enhance Corporate 142
Enhanced Oil Recovery 174
Enterprise Energy 84
Enterprise Migration Agreements 56
environmental impact statements 41–42, 47, 213
Environmental NGO Network on Carbon Capture and Sequestration 187
Epstein, Jonathan 142
Erbacher, John 52
Euromanga Basin, Qld 175
European Union 9, 175, 198
Evans, William 14
Excel Coal 83, 89
exchange rate *see* Australian dollar
'exclusives' 146
Exergen 142
exploration permits/licences 40, 43, 44–45, 48, 49, 83, 85, 217
exports (from Australia)
　black coal 4–5, 6, 8–10, 15, 29–30, 55, 56, 67, 147, 164, 197, 200, 204, 209–10, 213–16
　brown coal 58–59
　other industries 6, 31, 56, 201, 212, 213
Exxon Mobil 159

farmers 31, 36, 37, 43–45, 50–52, 92, 100, 101, 109–10, 112, 145–46, 197–98
Fairfax Media 72, 104, 147, 148–49
Felix Resources 82
Felton Downs, Qld 217
Ferguson, Martin 59, 92, 95, 162
fertiliser, coal-based 58, 191, 193
FIFO (fly in, fly out) workforces 52–54, 101, 111–13, 144–45
film festivals 124–25
fish 27, 56–57
Fitzroy River 59, 110
Flannery, Brian 82–83
Flannery, Tim 168, 169, 182
Flinders (explorer) 13
floods 101, 104, 107, 110, 183, 195, 197, 210
Flynn, Paul 86
Football Federation of Australia 117–18
Forbes' rich list 68
Ford 170
foreign guest workers 56, 94
foreign ownership 5, 34, 44, 98–99, 209, 214
Forrest, Andrew 2, 85
Fortescue Metals Group (FMG) 85
fossil fuel subsidies 133, 202

Franks, Scott 39
Friends of the Earth (FoE) 51–52
funding, philanthropic 75, 103–21
fund managers 65, 66–67
FutureGen project 180

G20 meeting (2009) 165
G20 membership 202
Galilee, Stephen 144
Galilee Basin, Qld 3, 6, 46, 55, 69, 72, 79, 84, 85, 90, 91, 93, 94, 96, 141, 175
Garnaut, Ross 164
Garnaut Climate Change Review xi, 12, 33
gas *see* coal gas; coal seam gas; 'natural gas'
gas companies 17, 18
gasworks 16, 18–19
Gazard, David 142
Geraldton, WA 14
Gillard, Julia 131, 183
Gillard government 94, 140, 192
Gippsland Basin, Vic 175, 176
Glasenberg, Ivan 62, 65–66, 68
Glencore (*later* GlencoreXstrata) 62, 66
Global CCS Institute (GCCSI) 165, 171, 184–87
global financial crisis 30, 88, 178, 198
global warming 31–33, 57, 75–76, 195–96, 204, 215
Gloucester Coal 123
Gold Coast United 117
Gollan, Robin 21
Goonyella Riverside mine, Qld 54
Gorgon LNG gas project 176, 180
government investors 65
Government of Singapore Investment Corporation (GIC) 65
Gowrie Junction, Qld 48
Graham, Geoff 128
grain 31, 44
grassroots campaigns 38, 40, 42, 48–52, 197–98, 217–18
Great Barrier Reef, Qld 3, 33, 56–58, 109, 197
greenhouse gases 3, 6–7, 9, 46, 58, 94, 102, 122, 162, 196 *see also* carbon dioxide; methane
Greenhouse Gas Research and Development Program 185
'greenhouse mafia' 8, 134–40
Greenpeace 64, 74, 151
Greens 8, 74, 130, 132, 183
Gregory mine, Qld 210
Greiner, Nick 142
Griffiths, Alan 141

INDEX   251

groundwater pollution 18, 47
Grudnoff, Matt 56
Gruen, David 212
guest workers 56, 94
Guildford Coal 141
Gulgong, NSW 109
Gunnedah, NSW 107, 109
Gunnedah Basin, NSW 76, 83
GVK 5, 70, 93–95
GVK/Hancock 70, 115, 143

Haggarty, Tony 83–84, 86
Halliburton 160
Hammond, Glennis 49
Hancock, Lang 69
Hancock Coal 60, 94, 95, 147
Hancock Prospecting 69, 147, 149, 150
Hansen, James 7
Hawke, Bob 135, 156
Hay Point, Qld 92
Hazelwood power station, Vic 58
health and safety issues
    coal mining 14, 19–24, 107–8
    see also air pollution; traffic problems; water pollution
Health and Social Harms of Mining in Local Communities (Sydney University)
heart disease 26, 27, 28
heatwaves 3–4, 7, 32
heavy metals 18
helicopter rescue services 106
Henry, Ken 131
Henry Lawson Festival 122
Herd, The (band) 125
heritage listings 39, 57, 59, 123, 197
Hillman, Ralph 129
Hinze, Russ 47
history
    coal mining ix–x, 13–15, 19–23
    mining companies 61–62
Hobart, Tas 17
Hooke, Mitch 131
horse breeding 39–40, 117
horse-drawn vehicles 24
housing scarcity 41, 53, 54
Howard, John 137–40, 141, 157, 160–62, 168
Howcroft, Russell 151
HRL Limited 86–87, 161, 180, 183, 217
HSBC 63, 64
Hunt, Greg 192
Hunter, Governor 15
Hunter Ports 116
Hunter River, NSW 14, 34

Hunter Thoroughbred Breeders' Association 39–40
Hunter Valley, NSW 15, 27–28, 30, 34–42, 86, 101, 105–6, 109, 115, 125, 144, 203, 213–14
Hunter Valley Protection Alliance 40–41
Hunter Valley Wine Industry Association 125
'hybrid' power stations 190
hydro electricity 25
hydrogen energy 172

ICAC 40, 42–43, 82, 83, 143
ice sheets 7, 195–96
ICF International 185
IMF 207
imports (to Australia) 6, 56
improper influence 143
Independent Commission Against Corruption (NSW)(ICAC) 40, 42–43, 82, 83, 143
India 9, 29, 55, 58, 59, 80, 90–93, 97, 190, 198–99, 207–8
Indigenous people 101–2
Indonesia 78, 208–9
industrialisation 25
infrastructure 46, 55, 70, 88–94, 98, 107, 159, 164, 175, 199, 206, 208, 218 see also ports, coal
insiders, political 135–36, 140–43
Institute of Public Affairs (IPA) 71, 150–53, 182
institutional investors 63–65, 66–67
insulation 201
Integrated Gasification Combined Cycle (IGCC) 189
Intergovernmental Panel on Climate Change (IPCC) 169–70
International Energy Agency (IEA) 158, 171, 180, 181, 185–86, 189, 190, 196
International Power 139
IPCC 169–70
Ipswich, Qld 14, 21, 48
iron ore 69, 72, 85, 212–13

Japan 9, 25, 29, 46, 130, 209
Jellinbah Group 82
Jellinbah Resources 61, 123
JFE Steel Corporation 81
jobs 55–56, 94, 128–30, 132, 212–13
John Hunter Children's Hospital 101
Jondaryan, Qld 49–50
Jones, Alan 217
Jones, David 4
journalists 146, 147–49

Joyce, Barnaby 95
JPMorgan Chase 63, 64

Kalpowar Aboriginal Land Trust 59
Karoly, David 4, 195
Keating government 136
Keephills project 187
Keeping, Brett 125
Kerin, John 156
Kevin's Corner mine, Qld 55
Keynes, John Maynard xi
Kianga mine, Qld 21
kidney disease 28
Kimberley region, WA 59, 218
Kim Yong, Jim 196
Kinghorn, John 86
Kininmonth, William 151
Kloppers, Marius 62
Kogan Creek Solar Boost project 190
Korea 9, 30, 46
Koukoulas, Stephen 216
Kurri Kurri, NSW 41
Kurrumbede station, NSW 44, 45
Kuwait Investment Authority (KIA) 65
Kwinana, Qld 172, 179
Kyoto conference 137, 157
Kyoto Protocol 140, 160–61

Labillardiere (explorer) 13
Labor Party 8, 40, 58, 129, 130–31, 139, 140, 142, 155, 162–63, 183–84, 192
labour camps 52–53, 54
Lake Liddell, NSW 28
Lake Vermont mine, Qld 82
Lander, Doris 49
Landscape Guardians 153
Lane, Greg 51
Latrobe Valley, Vic 12, 28, 58–59, 115, 133, 162–63, 203, 217
Laura Basin, Qld 73
Lawrence, Neil 130
Lawrence Creative 102
lead 19
Leigh Creek, SA 13, 14
LETAG 138, 161
Lewis, Essington 61
Liberal–National Party 40, 45, 46, 74, 75, 143, 155
Liberal Party 58–59, 129, 134, 140, 142, 151, 182
light bulb 24
lighting, public 16–18, 24, 25
lignite 58 *see also* brown coal

Linc Energy 60, 78–80
Lindsay, Peter 141
Lithgow, NSW 12
Liverpool Plains, NSW 43–45
lobbyists 63, 134–46, 171
Lock the Gate group 217
Lockyer, Darren 116
Logan, Francis 172
London, England 26
long wall mining 101, 125
Low Emissions Technology Demonstration Fund 161
Lower Emissions Technology Advisory Group (LETAG) 138, 161
Lowy, Frank 67
Loy Yang Power 153
lungs 22–24, 26–27, 28, 49

Macarthur Coal 48, 76, 81, 131
Macdonald, Ian 40, 43, 45, 83
Macfarlane, Ian 138, 162, 182, 192
McGuigan, John 86
McIndoe, Richard 133
Mackay, Qld 108
Mackellar, Dorothea 44
McMillan, William 20
Macmines Austasia 96–97
Madigan, John 153
Maharashtra, India 91
Maher, Tony 146, 167
Mallyon, Steve 87–89
Mandatory Renewable Energy Target 137–38, 140, 155, 204
Mangoola mine, NSW 37 *see also* Anvil Hill mine
Mannakal Economic Education Foundation 148
Mantle Mining 217
manufacturing industry 6, 13, 56, 213
Mao Tse Tung 75
Margaret River, WA 217
Mattson, Melissa 28
Maules Creek, NSW 76
MBD Energy 192
MCA 130, 131–32, 133, 135–36, 188
Meares, Anna 101
media 145–50
Meijin Energy 96–97
Melbourne, Vic 17, 18–19, 32
Melbourne Gas and Coke Company 17
mercury 26, 27, 28
Metallurgical Corporation of China 73
methane 20, 21, 146 *see also* coal seam gas; natural gas

Middlemount, Qld 76
Mineral Resource Rent Tax 126, 131–32, 133
Minerals Council of Australia (MCA) 130, 131–32, 133, 135–36, 188
Miners Accident Relief Act 21
Minewatch 38
mining equipment 86
Mitsubishi 120
Moatize 88
Monash Energy Project 172–73, 179
Monckton, Christopher 71, 148
Mongolia 85, 86
Moolarben mine, NSW 42
Moomba Carbon Storage 173, 179
Mooney, Tony 141
Moore, David 143
Moore, Norman 217
Moran, Alan 152–53, 182
Moranbah, Qld 28, 47, 52–54, 112, 113
Moranbah Medical Centre 54
Moranbah Traders' Association 54
Morgan, Hugh 151
Morgan Stanley 65
Mount Arthur mine, NSW 37
Mount Isa Mines 62
Mount Kembla, NSW 21
Mount Penny, NSW 83
Moura, Qld 21
Mozambique 87–88
Mudgee, NSW 42, 106–7, 109, 114–15
Muja power station 202
Mundra, India 90, 91
Munmorah Power Station 201
Murdoch, Rupert 151
Murray-Darling Basin 33
museums 122–24
Muswellbrook, NSW 41, 115
myth building 101–3, 126

Nahan, Mike 150–51
naphthalene 19
Napier, Ian 40
Napthine, Denis 151
National Australia Bank 63
National CCS Council 165, 181
National Low Emissions Coal Initiative 165
National Party 73
national security 33, 73
National Tourism Alliance 53
nation-building x
natural gas 18
Natural Resources Defence Council 187
Nebo, Qld 112
Nethercote, Ian 153

Network Ten 104, 147–48, 151, 218–19
Newcastle, NSW 12–14, 15, 25, 35, 77, 218
Newcastle Jets 116, 117
Newcastle Knights 116, 117
Newcastle Museum 122
NewGenCoal 102–3
New Hope Coal 47, 50, 61, 123
Newman, Campbell 58, 92, 197, 217
Newman, Maurice 153–54
Newman Government 74
New South Wales 12–13, 162, 197, 214–15, 216 *see also* Hunter Valley, NSW
newspapers 146–49, 151
Newstead Riverpark, Qld 19
New Zealand 85
Next Level Consulting Service, The 143
nitrogen oxides 26, 27, 28
Noble Group 142
noise pollution 107
Norges Bank 65
North Australia Project 71
Northern Australia and Then Some (Rinehart) 71
Northern Power Station 201–2
Northern Territory 203
Norway 65, 186
Norwich Park mine, Qld 210
nouveau riche 60–61, 67–81
Nova, Jo 148, 151
NSW Environmental Defenders Office 144
NSW Minerals Council 106, 124, 135, 144, 188, 205
nuclear power 162, 167
Nutall, Gordon 47–48

Obeid, Eddie 43, 83
O'Farrell, Barry 197
oil prices 29, 30, 193
O'Keeffe, Michael 87–89
open-cut mines 27, 30, 34–37, 45, 101, 123, 132
opinion polls 218–19
Oppenheimer, Nicky 66
Orica 125
Origin Energy 116, 173
Orwell, George 13
Otter, Nick 165
Otway Basin 172
'over-burden' 30, 35, 51
overseas workers 56, 94

Palmer, Clive 1–2, 5, 55, 56, 60, 62, 72–76, 80, 104, 107, 109, 112, 117, 142
Parer, Warwick 158

Parliamentary Inquiry (Moranbah) 54
particulate matter (PM) 26, 205–6
Passport Capital 88
Peabody 5, 39, 61, 62, 67, 83, 118–19
   community funding by 107, 109
Peak Downs mine, Qld 54
Pell, Cardinal George 151
Perth, WA 2–3, 17, 18, 32
PetroChina 84
philanthropy 75, 103–21
Pilbara iron ore mine, WA 212–13
pipelines, CCS 159, 173, 175–76
Pittsworth, Qld 109
planning 41–42
Playford B Power Station 202
Plimer, Ian 71, 150, 151
politics 128–45 see also major parties, by name,
   e.g. Labor Party
pollution permits 139
Porteous, Rose 69
Port Kembla, NSW 25
ports, coal 3, 30, 35, 46, 55, 56–57, 59, 91,
   92, 116, 218
power see electricity
PR campaigns 100–103, 119, 184, 186–87,
   191–93
PriceWaterhouseCoopers 196
Princeton University 170
private security companies 54
Productivity Commission Inquiry 157
professorships 118, 120
public relations see PR campaigns

QCoal 61, 81, 107
QR National 106 later Aurizon
Queensland 12, 30, 45–58, 59, 94, 163, 172,
   197, 214–15, 216
Queensland Department of Employment,
   Economic Development and Innovation
   47
Queensland Department of Mines 81
Queensland Gas 110, 119
Queensland Minerals and Energy Academy
   121
Queensland Nickel 142
Queensland Ombudsman 47
Queensland Resources Council 51, 121, 132,
   135, 143, 145
Queensland University of Technology 144–45

railways 15, 55, 164 see also Aurizon; trains
rainfall 32
Randolph, Marcus 180, 204

Ray, Greg 36–37
'Red Dog' 103
Reddy, G.V. Krishna 55, 93–95
Reddy, Sanjay 93
'rehabilitated' soil 38, 51
renewable energy see electricity, renewable;
   Mandatory Renewable Energy Target;
   wind power; hydro electricity; solar
   power
rents, housing 41, 53
research funding 118, 120, 121–22
Reserve Bank of Australia 29, 214
Reserve Bank of India 208
Resourcehouse Ltd 142
resource rent tax 126, 131–32, 133
Resources Super Profits Tax 2, 72, 131, 155
respiratory diseases 22–24, 26–27, 28, 49
Rey Resources 59
Rinehart, Gina 2–3, 5, 55, 60, 67, 68, 69–72,
   93, 95, 104, 142, 147, 151
Rio summit (1992) 7
Rio Tinto 5, 30, 62, 67, 76, 88, 89, 118,
   122–24, 131–32, 138, 150, 168, 172,
   210–13
   community funding by 105, 107–10,
      112–13, 115, 120, 126
Rising Tide 144
Riversdale II 89
Riversdale Mining 87–88
Rix's Creek, NSW 38
road problems 112–13
road trains 59
Robbins, Anthony 78
robotics 212–13
Roche, Michael 145
Rosemount winery 125
Roskam, John 151
Rowland PR 119
Royal Bank of Scotland (RBS) 64
Royal Commissions 20–21, 22–23
royalties, mining 35–36, 47, 58, 72, 74, 79,
   131, 214–15
Rudd, Kevin 2, 102, 128, 129, 130, 131,
   139, 140, 163–65, 184–85, 211
Rush, Martin 41
Rylstone, NSW 109

sacred sites 34, 39
safety see health and safety issues
Santos 107, 173
scholarships 120, 121
Scone, NSW 107–8, 113
sea level, rising 33
Seckold, Norman 85

Second World War 25
Secret Millionaire 79
Seeney, Jeff 143
Seven Holdings Group (SGH) 86
Shanxi province, China 86, 95–96
Shark Reef, Qld 57
Shell 84, 172
Shen Neng 1, MV 57
Shenua group 44, 107
Shortland, Lieutenant 13
shows, agricultural 109
Siemens 169
Sierra Club 208
Singapore 65
Singleton, John 149
Singleton, NSW 28
Sinodinos, Arthur 141
Smith, Bradley 51–52
soil 38, 44, 51, 85
solar power 137, 177, 190, 201, 204
South Australia 13, 173
South Galilee Project 85
South Korea 9 *see also* Korea
South West Hub 188–89
speculative investment 68, 69–70, 79, 80–81, 98, 207
sponsorship
    cultural 105, 122–24, 126
    educational 105, 118–22, 124
    sporting 101, 104–6, 116–18
    see also philanthropy
sports sponsorship 101, 104–6, 116–18
statesmanship x
steam engines 14
steamships 14, 15
steam trains 15, 25
steel-making 10, 13, 25, 29, 30, 46, 61, 69, 206–7
Stern, Lorraine 48
Stern, Nicholas 165, 170
St John, Edward 34
Stokes, Kerry 86–87
storage sites, CCS 175–76
streamflow 32 *see also* creeks
street lighting 16–18
strip-mining 34–59
strokes 26
subcritical coal technology 94, 189–90
subsidies, fossil fuel 133, 202
Sullivan, Greg 196
sulphur dioxide 26
Sumiseki 39
Sunnyside mine, NSW 210
supercritical coal technology 94, 189

super profits tax 2, 72, 131, 155
Surat Basin, Qld 46, 175
sustainability 95, 111, 115
Sustainable Minerals Institute (SMI) 118, 119, 122
Swan, Wayne 131
Swanbank Power Station, Qld 21, 201
Swan River, WA 18
Sydney, NSW 16–18, 19, 32

T4 proposal 218
Taiwan 9, 30
Tambling, Grant 139
Talbot, Ken 48
tar, coal 16, 18, 19
Tarong Power Station 201
Tarrant, Maria 139
Tarwyn Park 43
Tasmania 13, 194, 203
Tata Power 88, 169, 208
taxes 102, 130–34, 152, 214 *see also* carbon tax; resource rent tax; super profits tax
technology 24
television *see* ABC; Channel Seven; Channel Ten
temperature rises 31–33, 195–96, 204 *see also* global warming
Tenaska Trailblazer project 187
Teresa project, Qld 79
Thambimuthu, Kelly 181
Theodore, Qld 101
thermal coal 30, 68, 180–81
Thiess, Leslie 47
think tanks, right wing *see* Centre for Independent Studies; Institute of Public Affairs; Mannakal Economic Education Foundation
thoroughbred horse industry 39–40, 117
tiger conservation 91
Tinkler, Nathan 60, 62, 76–78, 84, 117–18
Titanic II 74, 76
tobacco 219
toluene 16, 19
Toowoomba, Qld 48–49, 109
Toowoomba Coal Mine Action Group 48
tourism 6, 31, 53, 56, 108–9, 125
town gas 13, 16–19
trace elements 38
traffic problems 112–13
trains 15, 24, 25, 107–8 *see also* railways
trams 24
Treadgold, Tim 149
Tropical Cyclone Oswald 195
Turnbull, Malcolm 162
turtle program 110

Tyrrell, Bruce 40, 125
Tyrrells Wines 40

UBS 64–65
UCG 78–79, 80
Ulan mine, NSW 42
UN see United Nations...
underground coal gasification technology see UCG
United Nations Clean Development Mechanism (carbon credits) 92, 94, 178–79
United Nations climate negotiations 136 see also Kyoto conference
United Nations Educational, Scientific and Cultural Organization (UNESCO) 57–58, 197
United States 9, 23–24, 29, 80, 160–61, 180, 198, 208
University of Central Queensland 121
University of Newcastle 120–21
University of NSW 120, 121
University of Queensland 118, 119–20, 121–22
University of Sydney 120
University of Wollongong 120, 121
Utah mining company 29, 54

Vaile, Mark 141
Van Diemen's Land (later Tasmania) 13
Verrender, Ian 132–33
vertosols 44, 51
Victoria 12, 58–59, 162–63, 172–73, 196, 214–15 see also brown coal; Latrobe Valley, Vic
Vietheer, Mark 50

wage levels 41
Wallerawang Power Station 201
Wallin, Chris 80–81
Wambo Coal 39
Wambo Homestead 39
Wandoan, Qld 46, 51, 66, 179–80
war and steel-making 25
Waratah Coal 1, 60, 112, 213
Warkworth mine, NSW 124
water pollution 18, 42, 47, 107, 110

water resources 32, 101, 110, 176–77, 196, 207
'wedges' approach 170
Wells, Dick 181–82
Wells, Kim 196
Wesfarmers 123
Western Australia 13, 59, 172
WesTrac 86–87, 116
wheat 44
White Energy 82–83, 86
Whitehaven Coal 60, 61, 76, 77–78, 83, 84, 109, 141, 210
Wiles, Andy 40
Williams, John 41–42
Williams, Nikki 128, 144
Wilpinjong mine, NSW 42
Wiltshire, Jim 48
wind power 137, 138, 153–54, 190, 203, 204
winemakers 40–41, 125
Wolfensohn, James 165
Wollar, NSW 42
Wollongong, NSW 12
Wongai project, Qld 59
Woods Mackenzie 84
workers' compensation 21, 22, 23
World Bank 178, 199, 208
World Health Organization 205
World Heritage listing 57, 197
World Resources Institute 187
Wuhan Iron and Steel 88
WWF-Australia 167

Xenophon, Nick 154
Xstrata (later GlencoreXstrata) 5, 30, 37, 51, 61, 62, 64, 65–66, 67, 131, 139, 142
 community funding by 101, 104, 105, 106–7, 110, 111, 113, 118, 120, 122, 126

Yabulu Nickel Refinery 75
Yancoal 38, 61, 82, 122, 123
Yao Jinli 97
Yao Junliang 95–97
Yarrowmere, Qld 96

ZeroGen project 172, 180, 181

www.ingramcontent.com/pod-product-compliance
Lightning Source LLC
Chambersburg PA
CBHW031725230426
43669CB00007B/245